The Maya Forest Waterlands

This book examines the entanglements and blurred edges of nature conservation and geopolitical relations in the borderlands of the trinational Maya Forest.

Maya Forest is an umbrella term for transboundary conservation developed by scientists and conservationists in the 1990s to protect the threatened rainforest in the borderlands of Mexico, Belize, and Guatemala. Currently, the Maya Forest is a biodiversity hotspot composed of a network of protected areas and heritage sites. However, issues related to water, land, and forests have often been treated as separate political units, and not as part of the same history. Written by two authors with decades of hands-on experience in this region, this book sheds light on the complex dynamics by which conservation and natural resource management geopolitically shape borderlands such as the Maya Forest. The book introduces the novel concept of forest waterlands as borderlands and fluid edges, which are now subject to concern by conservationists. These are entangled spaces in which conservation, people, and politics interact, connect, and disconnect with the nexus of waters, forests, and lands. The book sheds light on the building and mapping of the Maya Forest ecoregion, with particular attention to water as an often neglected, but unifying element. It showcases how the Maya Forest is a distinct region characterized by transformations entangled with the Maya, trails of biological stations, the shared history of *chicleros* (chewing-gum hunters), fluid international rivers and transboundary basins, and various geopolitical discrepancies. It offers a contemporary glimpse into the Maya Forest's intertwined bio- and geopolitics, which urge us to rethink borders and boundaries.

This book will be of great interest to students and scholars of nature conservation, global environmental politics, geopolitics, borderlands, international relations, and natural resource management.

Hanna Laako is a senior researcher at the University of Eastern Finland (UEF). She worked as a researcher in southern Mexico for ten years and currently leads a research project on Political Forests – the Maya Forest – financed by the Mexican Council of Science and Technology CONACYT and the Finnish Kone Foundation. She holds a PhD in Political Science and International Relations from the University of Helsinki, Finland.

Edith Kauffer is a senior researcher at CIESAS (Centro de Investigaciones y Estudios Superiores en Antropología Social) in Mexico. She is the former coordinator of the researchers' water network on the border between Mexico, Guatemala, and Belize (RISAF) (2003–2016) and the co-coordinator of the Comparative Research on Regional Integration and Social Cohesion Consortium (RISC) working group on Management of Strategic Resources, Environment, and Society. She holds a PhD in Political Science from Aix-Marseille University, France.

Routledge Studies in Conservation and the Environment

This series includes a wide range of inter-disciplinary approaches to conservation and the environment, integrating perspectives from both social and natural sciences. Topics include, but are not limited to, development, environmental policy and politics, ecosystem change, natural resources (including land, water, oceans and forests), security, wildlife, protected areas, tourism, human-wildlife conflict, agriculture, economics, law and climate change.

Conservation Concepts
Rethinking Human-Nature Relationships
Kurt Jax

Japan's Withdrawal from International Whaling Regulation
Edited by Nikolas Sellheim and Joji Morishita

Conservation Leadership
A Practical Guide
Simon Black

Positive Psychology and Biodiversity Conservation
Health, Wellbeing and Pro-Environmental Action
Jolanta Burke, Sean Corrigan, Jimmy O'Keeffe, and Darren Clark

Conservation in the Anthropocene
Reshaping Interaction with Nature
Fred Van Dyke

The Maya Forest Waterlands
Shared Conservation, Entangled Politics, and Fluid Borders
Hanna Laako and Edith Kauffer

For more information about this series, please visit: www.routledge.com/Routledge-Studies-in-Conservation-and-the-Environment/book-series/RSICE

The Maya Forest Waterlands

Shared Conservation, Entangled Politics, and Fluid Borders

Hanna Laako and Edith Kauffer

Routledge
Taylor & Francis Group
LONDON AND NEW YORK

earthscan
from Routledge

First published 2025
by Routledge
4 Park Square, Milton Park, Abingdon, Oxon OX14 4RN

and by Routledge
605 Third Avenue, New York, NY 10158

Routledge is an imprint of the Taylor & Francis Group, an informa business

The authors acknowledge funding support from The University of Eastern Finland to enable open access distribution of the book.

British Library Cataloguing-in-Publication Data
A catalogue record for this book is available from the British Library

ISBN: 978-1-032-54930-9 (hbk)
ISBN: 978-1-032-55111-1 (pbk)
ISBN: 978-1-003-42905-0 (ebk)

DOI: 10.4324/9781003429050

Typeset in Sabon
by Taylor & Francis Books

Contents

Illustrations

Images

Figures

Maps

Tables

About the Authors

Hanna Laako holds a PhD in Political Science and International Relations from the University of Helsinki, Finland (2011), and a title of docent (adjunct professor) in Social and Public Policy, especially International Relations, from the University of Eastern Finland (2022). Currently she is a senior researcher at the Department of Social Sciences of the University of Eastern Finland (UEF), conducting research as part of the Borders, Mobilities, and Cultural Encounters Research Community (BOMOCULT). She worked as a researcher in southern Mexico for ten years, including for different research institutions such as *El Colegio de la Frontera Sur* (ECOSUR), *Centro de Investigaciones y Estudios Superiores en Antropología Social* (CIESAS-Sureste), *El Centro del Cambio Global y la Sustentabilidad* (CCGS), and *El Centro de Investigaciones Multidisciplinarias sobre Chiapas y la Frontera Sur* (CIMSUR-UNAM). She pertained to the Mexican National System of Researchers until the end of 2021. At the moment, she is leading her research project on *Political Forests – the Maya Forest* – funded by the Mexican Council of Science and Technology CONACYT (2019–2020) and the Finnish Kone Foundation (2020–2025). She has previously been a PI of several research projects dealing with borderlands and midwifery, among others. Her fields of expertise include International Relations and Political Geography, Borderlands Studies, Conservation Politics, Movements, and Decolonization. During her career, she has been involved in many networks and collaborated with different organizations, stakeholders, and rightsholders, for example, those of Indigenous people, midwives, photographers, and conservationists. She has conducted extensive fieldwork in many southern Mexican states, Belize, and Guatemala. https://orcid.org/0000-0003-0124-0711

Edith Kauffer holds a PhD in Political Science (Aix-Marseille University, 1997). Currently she is a senior researcher at CIESAS (*Centro de Investigaciones y Estudios Superiores en Antropología Social*) in Mexico. She is the former coordinator of the researchers' water network on the border between Mexico, Guatemala, and Belize (RISAF) (2003–2016) and she is currently the co-coordinator of the Comparative Research on Regional Integration and

Social Cohesion Consortium (RISC) working group on Management of Strategic Resources, Environment, and Society. She is also associate editor of the Journals *Regions & Cohesion* (Berghahn Journals), *Sociedad y Ambiente* (Mexico) and belongs to the board of the *International Journal of Water Resources Development* (Taylor and Francis) and *Íconos Revista de Ciencias Sociales* (FLACSO – *Facultad Latinoamericana de Ciencias Sociales* –Ecuador). She conducts research on transboundary waters and river basins, water policy and politics, gender and water, and political and anthropological dimensions of sediments. During the past three decades of fieldwork in the southern Mexican and Central American borderlands, she has been the P.I. of 33 projects. She has also received funding for developing training with NGOs, grassroot organizations, and government stakeholders related to water issues for whom she developed proposals for river basin programs. Throughout Central America, she has been advising and teaching on water issues for regional cooperation, for example for the Central American Commission on Environment and Development (CCAD in Spanish). Over three decades, she has witnessed deep transformations in the forest waterlands and proposed, as the P.I. of a Mexican team, the first delimitation of the six transboundary river basins on the borderlands of Mexico, Guatemala, and Belize, at a time when the governments of these states were not interested in the topic (2004–2011). She participated in river basin councils and committees in Mexico to elaborate a proposal for the new general water law in Mexico and organized numerous international meetings on borders and waters in southern Mexico and Central America. https://orcid.org/0000-0003-4238-9782

Acknowledgements

During the past decades, many conservationists have promoted rewilding as an approach to let nature take care of itself, to repair damaged landscapes and ecosystems, and to allow natural processes to shape more biodiverse habitats. This book is a kind of rewilding of social and political sciences, based on the transboundary concept of the Maya Forest, but accompanied by *waterlands*, which has invited us to extend beyond our accustomed categories, allowing them to take us to their blurry edges and interstices. In the book, we revisit many locations in the Maya Forest Waterlands, exploring their shared conservation approaches, entangled politics, and fluid borders, intending to listen, and if we are lucky, to transmit some of those rewilded nexus. In this way, we seek to capture the many faces, shared histories, and contemporary transformations of a transboundary region, which has also been defined as a biodiversity hotspot and a basis for Mayanism at a scientific crossroads.

This book is also a result of many other scientific crossroads and encounters. One key juncture is definitely our decision to seek to be published in the *Routledge Series on Conservation and Environment*, whose early publications motivated us to propose this title. The series is dedicated to inter-disciplinary approaches and to integrating perspectives from both social and natural sciences. While our book is first and foremost grounded in the social and political sciences to which we pertain; given the topic, it is deeply connected to natural sciences and multi-disciplinary approaches. We thank the series editors and reviewers for their helpful feedback and support.

This book is a crossroads and encounter between us, the two authors, who have collaborated on topics related to borders, waters, and forest conservation since 2017. The book began to take shape in the fall of 2022, and it brings together our long-term trajectories, our shared fieldwork between 2017 and 2019, and both individual and shared research results gathered for this book since 2019. Hanna Laako would like to extend her sincere gratitude to Edith Kauffer for this intensive adventure and important learning experience. Together, they also thank their colleagues, especially those from the shared research project of the Usumacinta River basin (2017–2019). In particular, our warm thanks to Miguel Ángel Urbina Peréz and Esmeralda Pliego Alvarado, who accompanied us to many fieldwork locations and shared insights over the years.

We also jointly extend our gratitude to Pavel Popoca Cruz from the *Instituto de Ecología-UNAM*, who kindly elaborated the maps for this book.

In addition, Hanna Laako is indebted to the many conservationists, NGOs, community members, and institutional representatives who have taken part in this transboundary research on the Maya Forest since 2019. These organizations and people have allowed her to learn not only about conservation work, but also about the complex social, territorial, and (geo)political dynamics that characterize the Maya Forest today. She has been delighted and inspired by evidencing active civil society participation, critical thinking, transboundary collaboration, and alliance building. There have been plenty of challenges, and Laako truly appreciates the time and space these many stakeholders and rights-holders have given to enable her to conduct this research. It would have not been possible without these insights.

She is grateful for the many academic institutions where she has previously worked in Mexico and with which she continues to engage, especially *El Colegio de la Frontera Sur* (ECOSUR), *Centro de Investigaciones y Estudios Superiores en Antropología Social* (CIESAS), and *Centro del Cambio Global y la Sustentabilidad* (CCGS). The Mexican Council of Science and Technology (*CONACYT*, now *CONAHCYT*) supported Laako's research on the Maya Forest between 2019 and 2020. Laako also extends sincere thanks to the *Institute for Social and Cultural Research* of the *National Institute of Culture and History* in Belize and its knowledgeable personnel for the extraordinary support in terms of research in Belize. She also extends warm thanks to the *Belize Maya Forest Trust* and the *Friends for Conservation and Development* for the encounters, long discussions, collaboration, and support. Her gratitude extends to such important bodies as *Asociación Balam* and several colleagues at the *Centro de Estudios Conservacionistas* (CECON) in Guatemala and *Natura Mexicana* in Mexico. A special thank you to Miguel Ángel Urbina Peréz for accompanying her during fieldwork across the Maya Forest between 2019 and 2022 and for the many significant insights. Thank you to Miguel Díaz Perera, who initially provided reading suggestions to deepen the history of *chicleros*. There are many others to thank, who are not mentioned by name to protect anonymity – your contributions are greatly valued and not forgotten. Possible errors or misinterpretations are Laako's sole responsibility.

In 2020, Hanna Laako moved back to her country of origin, Finland, leaving southern Mexico, where she had been living for the previous decade. In Finland, she is particularly indebted to the *Kone Foundation*, which awarded her a grant (number 201902764) for the Maya Forest research conducted between 2020 and 2025. Moreover, the Finnish Foundation *Kansan Sivistysrahasto* and the *Faculty of Social Sciences and Business Studies* of the UEF have provided her with a small grant and an honorarium to enable transboundary fieldwork, for which she is grateful. She would also like to extend her gratitude to the *Department of Geographical and Historical Studies* of the UEF, which strongly supported the fully open access route for this book, as well as the fieldwork. She warmly thanks the *Department of Social Sciences* of the UEF, which provided unconditional support for the writing process and language revisions. At the UEF, she is especially indebted to Irmeli Mustalahti and Antti Erkkilä from the Responsive Natural Resource Group (RNRG), Ari Lehtinen, and many colleagues at the Environmental Policy group, and Tiina Sotkasiira, Vadim Romashov, Timo Toikko, and colleagues from the Department of Social Sciences. Many thanks also to the *Borders, Mobilities and Cultural Encounters Research Group* (BOMOCULT), where distinct border-related issues have been shared. In particular, Laako extends her gratitude to the collaborators of the BOMOCULT seminar "(De)Colonizations in Global Borderlands" (organized on May 15, 2024), and the writers' workshop, organized together with Irmeli Mustalahti in May 2024. During both events, which involved Kauffer's visit to Finland, findings related to this manuscript could be discussed and advanced.

Last but certainly not least, Hanna Laako extends her gratitude to her closest family and friends for having her back, for accompanying both storms and sun, for the always nurturing and insightful discussions and sharing of daily life: Katariina Kantola, with whom she talked about *chicleros*, ¡kiitos Sis!; Marisol Jiménez, with whom she traveled to the Maya Forest and shared insights, ¡te quiero mucho amiga!; and Margarita Barajas, with whom she visited the rainforest and wildly debated its many transformations, ¡gracias infinitas amiga!. To Laako's extended Mexican family, who have been there for her an endless number of times, and her own parents, who also provided crucial logistical support to enable long fieldwork trips and travel. To Manuel Martínez Espinoza, with whom she always shared the "podcasts" and "master classes" about academic endeavors and research paths. And finally, to Laako's own quirky tribe: Her mom, her brother, and her two fabulous girls, with so much love, gratitude, and admiration. Hanna Laako dedicates this book to her father, Tero Laako, who unexpectedly passed away on a freezing cold January night of 2024 while this manuscript was being written and who instilled in her the passion for photography, science, and nature. *In warm memory of my father.*

Edith Kauffer deeply thanks Hanna Laako's invitation, leadership, and insights in this collaborative adventure, which was an opportunity to rethink three decades of fieldwork in Mexico, Guatemala, and Belize. This book required some jumps into the past to re-signify research experiences and update data as well as discuss old and new highlights, sharing and confronting perspectives.

Edith Kauffer's fieldwork would not have been possible without numerous fundings, truth, and collaborations with colleagues, among which she would like to mention the VAL-USES project (2018–2022), funded by both the *Agence Nationale de la Recherche of France* (ANR-17-CE03–0012–01) and the *Consejo Nacional de Ciencia y Tecnología* of Mexico (Mexican Council of Science and Technology) (FONCICYT-290792). Kauffer is indebted to the riverine people living in the Usumacinta River basin in Mexico and Guatemala whom she interviewed, sand miners, numerous colleagues, and research collaborators – especially Pierre Charruau and Isabelle Michallet, the Mexican and French PIs. In particular, she thanks Víctor Alfonso Gallardo Zavaleta, Sofía Ruiz Díaz, and, above all, Miguel Ángel Urbina Perez, who shared their valuable contribution during extended fieldwork navigating the numerous rivers of the Maya Forest waterlands and going ashore in 79 sedimentary areas or islands.

Concerning the FORDECyT-project 273646 approved by the Mexican Council of Science and Technology (2017–2019), she extends her sincere gratitude to all team members involved in the above-mentioned fieldwork, especially the social science group she coordinated and colleagues from Guatemalan institutions – particularly Luzma Fabiola Nava (CCGS) and José Alberto Gallardo Cruz (Universidad Iberoamericana), general PIs. Fieldwork and information for Chapter 4 were made possible by the objective JJ of RTMG project number 291987, funded by the Mexican Council of Science and Technology (2017–2018), and the water security axis coordinated by Kauffer, including David Andrade García and Ludivina Mejía González, to whom she is grateful. Fieldwork at the Hondo River in Mexico was carried out as part of project PDCPN 2004 number 248954 in 2014–2018, funded by the Mexican Council of Science and Technology. For the fieldwork in Belize (2014), which was eventually funded by RISAF's project number 244120, she thanks the Mexican Council of Science and Technology and Luisa Fernanda Thompson Poo for her administrative assistance and events organization.

Edith Kauffer would also like to thank her own research institution, the *Centro de Investigaciones y Estudios Superiores en Antropología Social*, for the opportunity to freely develop her research over 18 years as well as the possibility to enter into partnerships with other Mexican institutes where she carried out the above-mentioned projects: *Centro del Cambio Global y la Sustentabilidad* (CCGS), *El Colegio de la Frontera Sur* (ECOSUR), and *Centro de Investigación en Ciencias de Información Geospacial* (Centro Geo). She also recognizes the hard and dedicated work of CIESAS Sureste's administrative team. Additionally, she extends her acknowledgments to the University of Eastern Finland (UEF) and the Kone Foundation, which funded her visit to Finland in May 2024 to participate in the "(De)Colonizations in Global Borderlands" seminar, and to Irmeli Mustalahti for her invitation to the writers' workshop 2024.

She extends her gratitude to Víctor Alfonso Gallardo Zavaleta and Rubén Darío Gutiérrez Campo, both Ph.D. students, for the rich debates about Mexican water policies, Colombian literature on the Magdalena River, and sedimentary issues, which have enriched the waterlands concept, as well to José Luis Escalona Victoria for his literature suggestions to better understand the "Mayas", which have stimulated the "Mayan rivers" discussion.

Introduction

Hanna Laako and Edith Kauffer

Conservation, understood as a human practice to protect nature and safeguard biodiversity, has played an important role in numerous territorial and regional transformations around the world. It has contributed to questioning many assumptions related to both natural and *social sciences* – for example, the Anthropocene with its challenging human/nature relationships. It is involved in political borderings and impacts how we perceive the so-called peripheries, many of which are now identified as biodiversity hotspots subject to environmental concern. Often, different scholars, stakeholders, and actors have set out for collaboration based on an arduous attempt of coming-together in deeper mutual understandings to advance conservation, societies, and the sciences. Occasionally, the same endeavor has also evoked old conflicts and unfolded new ones. Even shared conservation is intimately entangled with politics and borders, both epistemologically and empirically.

This book, *The Maya Forest Waterlands*, emerges from these deep entanglements to critically explore the ways in which conservation, politics, and borders connect and disconnect both in the interstices of natural and social sciences and, empirically, in a transboundary region identified as a biodiversity hotspot. This is the Maya Forest, located in the borderlands of Belize, Guatemala, and Mexico. Beyond its contemporary trinational political borders, these are also the territories of the ancient Mayan civilization and homelands to many people who self-identify as Mayas or as Indigenous. Many other inhabitants and settlers who also pertain to the contemporary Maya Forest landscapes and shapeshifting realities do not identify as such.

These lands have been subject to considerable territorial, political, and social changes during the past decades, particularly related to such multifaceted – and simultaneous – phenomena as tourism and drug-cartel activity. While these lines are being written, a major railway infrastructure project, the *Maya Train*, is being inaugurated in the Mexican Yucatán Peninsula, urging territorial changes and stirring up debates over the developments of the "long forgotten and abandoned southern Mexico", with initial plans to expand to Belize and Guatemala. At the same time, the drug cartels, with their changing geographies, are expanding across all the border areas of the region, causing violence and fear by extorting resources and land from people. These are only two examples of

DOI: 10.4324/9781003429050-1

complex, simultaneous, and even contradictory entanglements, and they deeply affect the region's natural environments, which are formed today around a broad web of protected areas, conservation activities, and natural resource politics.

Scientists and conservationists introduced the concept of the Maya Forest in the 1990s to protect the humid tropical rainforest located in the borderlands of Belize, Guatemala, and Mexico (Laako et al. 2022). The Maya Forest region is not fixed but fluid, depending on who is using the term and to what ends. Usually, however, it tends to include the whole of Belize, the department of Petén in Guatemala, and the Mexican states of Chiapas, Tabasco, Campeche, Quintana Roo, and Yucatán. An alliance of different local, national, regional, and international organizations – environmental organizations, non-governmental organizations (NGOs), and governmental institutions – sought to safeguard the transboundary continuity of the largest Mesoamerican rainforest and, in doing so, generate multidisciplinary scientific collaboration, enhance connectivity between protected areas and across political borders, further (eco-) tourist revenue, and form a shared conservation space. It was based on the understanding that the region shared common historical (mainly Mayan) roots and ruins, and was united in advancing future aspirations related to nature conservation, hampered by the existing and dividing political borders and policies concerning natural resources.

Instead of silently fading away with the passing of time, the concept of the Maya Forest has outlived its initial introduction, inciting many competing mappings and contested discourses. In terms of territory, the Maya Forest is important. According to the World Bank (2022), over 35% of Belizean lands are protected, while the areas protected in Mexico have increased to nearly 15%, and the Guatemalan rate currently stands at around 20%. The numbers exclude marine protected areas, which have also been increasing. These areas are presented in Map 0.1, which illustrates both the Maya Forest region and its broad web of protected areas.

Many of the Maya Forest's protected areas are located along the borders and include international categories such as Ramsar wetland sites, UNESCO natural heritage sites, and biosphere reserves, which importantly connect the Maya Forest to global conservation geopolitics. Other categories are national parks, wildlife protection units, as well as federal, departmental, state and municipal ones. The protected areas also include numerous private and voluntarily protected areas, many of which are managed by communities, each with a unique history and current tendencies. These, again, interact with other human activities and developments.

However, the Maya Forest (*Selva Maya* in Spanish) is not only the sum of its protected areas. At least two contemporary megaprojects are worth mentioning: first, the trinational conservation project *Selva Maya Programme* (https://selvamaya.info/en/mayan-rainforest/), which has been actively working in these borderlands for the past decades, promoting transboundary conservation; second, the new megaproject *Five Great Forests of Mesoamerica* (https://www.rewild.org/get-to-know/five-great-forests-of-mesoamerica), which includes the Maya

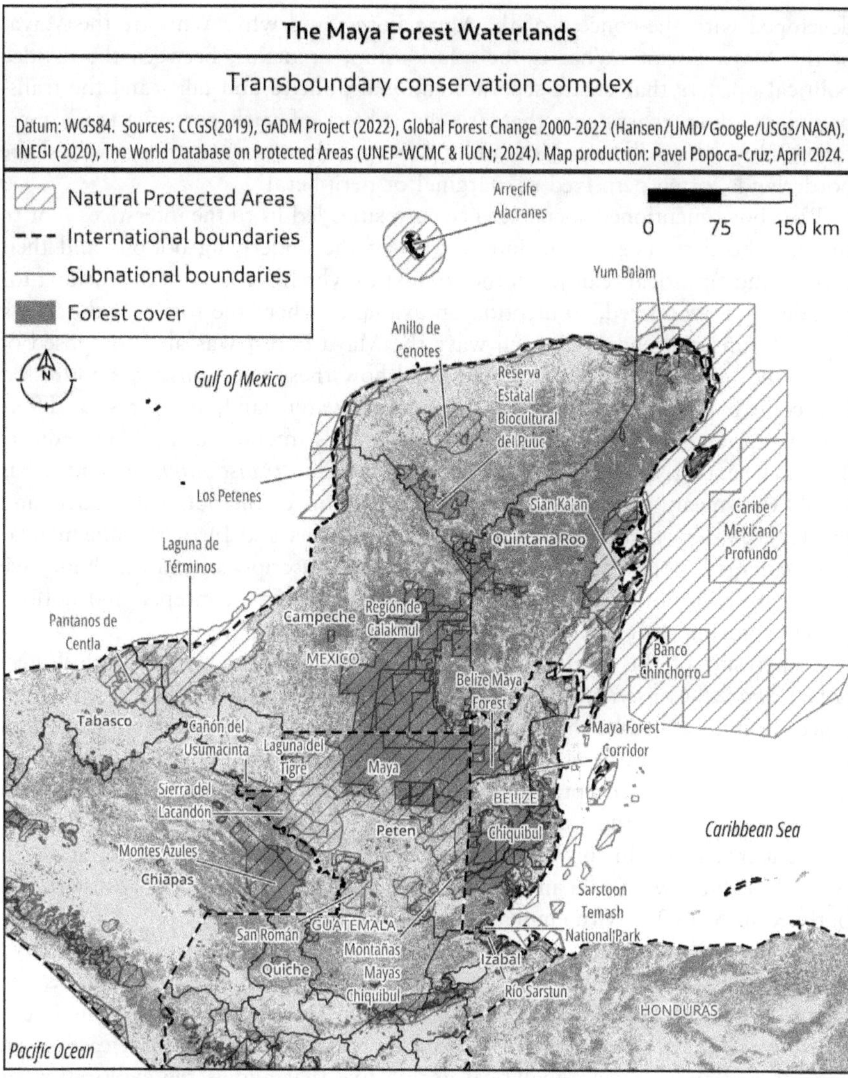

Map 0.1 Maya Forest Transboundary Conservation Complex
Source: Pavel Popoca Cruz

Forest. The latter, in particular, is intimately connected to borders, as all the five great forests identified by the project are directly located along and across them. These megaprojects emerge from the earlier intergovernmental Mesoamerican Biological Corridor.

As we are political scientists working on issues related to borders and nature, the above-mentioned developments in and of the Maya Forest incited us to ponder, among others, the following questions: who are the actors that developed with the concept of the Maya Forest and why? Who are the Mayas of the Maya Forest? What is the relationship, or nexus, between the divided political borders that create separate national policies and units and the transboundary, fluid ecosystems that are the subject of such conservation aspirations? How does this sort of ambitious, transboundary conservation shape borderlands, often perceived as marginal or peripheral?

The above-mentioned social-science curiosities led us to the interstices – or to the very borders, nexus, and interfaces – of the underlying notions and their contrasting empirical realities in the context in which we have been working for decades. We wondered, to mention an example, where the forest of the Maya Forest begins and ends, in which ways the Maya Forest was also composed of waterlands, river basins, or sinkholes, and how these questions were addressed in a context where national legislations divide water, land, and forest in different categories. How do waters, forests, and lands interact in the Maya Forest? To what extent was transboundary conservation *transboundary*, and what would that mean, in practice, in different locations of the political borders and the very borders themselves? Many other questions and further problematizations emerged along the way of writing this manuscript, and in this book, we constantly zoom in and out of many given assumptions, concepts, and notions while exploring their interstices and entanglements.

As a result of this challenging endeavor, here in *The Maya Forest Waterlands: Shared Conservation, Entangled Politics and Fluid Borders*, our main objective is to elucidate the shared and disconnected waters, lands, and forests while exploring how conservation shapes borderlands. For this purpose, we introduce a novel concept: *forest waterlands*, which are both inhabited and uninhabited borderlands subject to ecological concern that manifest in many different interfaces. In this book, we refer to the "Maya Forest waterlands" in two ways. First, we use it in capital letters to refer to the Maya Forest region outlined in Map 0.1, with emphasis on the fact that this is not only a forest but also waterlands. Second, we refer to the Maya Forest waterlands (or Maya forest waterlands in lower case), as an analytical perspective that places the nexus of Maya/forest/water/lands under critical scrutiny. For us, in this book, forest waterlands are not given or fixed ecological or geographical categories, such as wetlands, but an analytical perspective that allows the exploration of these sociopolitical constructs. In a similar vein, international borders are human constructs and therefore always subject to politics, while the landscapes they intend to control, as borders, are active with their own life. While political borders are often considered fixed in natural elements, typically drawn in rivers or mountain ranges, they are eventually "fluid", that is, not only liquid but entailing intertwined society/nature relationships and borderlands that are unstable, intermittent, vague, and subject to change. Those fluid borders, which tend to be on the edges of our understanding, scientific or other knowledges, are also great places for learning, encounter, and connectivity.

Therefore, in this book, we argue that the Maya Forest waterlands involve entangled spaces where conservation and natural resource management interact, connect, and disconnect with the nexus of waters, forests, and lands. In this book, we shed light on the building of the Maya Forest ecoregion, with particular attention to water as an often neglected but unifying element. The book portrays blurred edges and open-ended encounters between Maya Forest inhabitants, conservation, forests, waters, and lands. At the same time, it offers a contemporary glimpse into the Maya Forest's political geographies, international relations, and geopolitics, which are today characterized by the existing multiple borders: those between countries, states, and municipalities as well as around the protected areas that align to the former ones.

This introduction continues by presenting the many mappings of the Maya Forest, first by outlining the formation of both political and protected area borders (Map 0.1) and then by gearing toward the three connecting, transboundary socioenvironmental corridors identified within the Maya Forest. These corridors allow us to analyze the region from a distinct angle, and we use them throughout the book to structure our findings and writing. We then continue by mapping the Maya Forest waterlands as a science space and place, and the existing studies we build on and reflect upon. At the end, we describe the research we have conducted for this book, which materializes in the structure of the book with the continuing chapters.

Mapping the Maya Forest Waterlands: Bordered Lands, Forest Walls, and Waterlines

As Map 0.1 shows, the Maya Forest Waterlands are formed of many borders, which are, again, a result of multiple mappings in different times and places. In fact, as this book will show in its many pages that will follow, an important characteristic of the Maya Forest is that it consists of rigorous, overlapping, and competing mappings, both prior and posterior to the building of this particular concept in the 1990s. Some of these mappings correspond to the formation of national borders and those of the countries' inner states, municipalities, and localities. While nature and natural elements are always implicated in these political borders, other mappings particularly focus on natural resources. These include those of conservation, materialized in protected areas, many of which are located along political borders. Harris and Hazen (2015) have called this "mapping for conservation" and "conservation cartographies", which also characterize the Maya Forest origins and current developments. In this book, our objective is to interrogate these mappings as power relations entwined with the other mappings pertaining to this complex region.

Mapping has always been intimately involved with power relations, (geo) politics, and coloniality. It binds together and separates natural environments, people, and societies. Mapping is also embedded in knowledge production, and thus, both natural and social sciences play a role in what and who is mapping and for what purpose. Additionally, mapping is always open-ended with respect to how it will be eventually used. Mapping affects our ways of perceiving and acting in the world. For this reason, many environmental historians, for example, have used the borderlands perspective to rewrite many national histories, and in doing so, they have remapped the dominant narratives, expanding toward global history that has allowed centering other peoples and different environmental boundaries, such as those of the Indigenous or of oceans (e.g., Metcalf, 2020; Prado, 2012).

According to Harris and Hazen (2015), the "power of maps" is always related to representations, positionality, and partiality of knowledge, including their silences and absences. These challenges of mapping are well illustrated in recent academic debates concerning the coloniality of mapping and the ways in which coloniality has been questioned by critical scholars (e.g., Lucchesi, 2018; Wainwright, 2008). Alternative methods, such as counter-mapping and participatory mapping, have been introduced (e.g., McGurk & Caquard, 2020; Sletto et al., 2020). The latter intends to benefit different communities, people, and the Indigenous in their struggles for land, territory, and natural resource rights often left out of official state mapmaking. Yet, occasionally, further problems emerge from giving away information concerning delicate resources that may be exploited by unexpected parties, such as biodiversity pertaining to collective Indigenous knowledges. Indeed, and as this book will also suggest, in some occasions, both popular and/or political silence is extended to public mapping to avoid conflict and exploitation (e.g., Rahder, 2020). Not all silences and absences are negative nor all inclusions positive; thus, mapping and *unmapping* involve both revelatory elements and unexpected results.

Mapping always involves space and place: the space, which is digital or on paper, and the place, which involves a lived-in, subjective location. The mapping space may refer both to the end result on paper/in digital form and to the contemporary Geographic Information Systems (GIS) work that usually takes place within offices and computers, possibly far away from the actual location subject to mapping. The mapping place, again, refers to the physical location or route in reality, which is always lived-in and subjective. As well described by Rahder (2020) in the case of the Maya Biosphere Reserve, contemporary conservationists divide their time between these two: between the "remote" field (*campo*) away from the urban or power centers as a mapping place, and the "remote" offices and computers often distant from the field as the mapping space. Ingold (2000) has gone even deeper in dividing the concepts of *mapping* and *mapmaking* in these respects. For him, *mapping* is closer to the physical place. It is wayfinding and storytelling based on bodily motion in a territory that creates histories, not just maps. Again, *mapmaking*, for him, is divorced from the need to travel, eradicating the movement, stories, and bodies and resulting in an abstract map, a sort of cartographic illusion that conveys the perception of well-defined borders and control.

In this section, we continue by first zooming into what looks like the well-defined borders of the Maya Forest Waterlands in Map 0.1, namely, international borders and protected areas. We examine the main mapping and mapmaking processes to establish the point of departure for our book's main concerns. To do this, we take the reader for a walk across time and place in Map 0.1, although we also challenge the reader by shifting the angle to gain a broader, strategic perspective (Map 0.2).

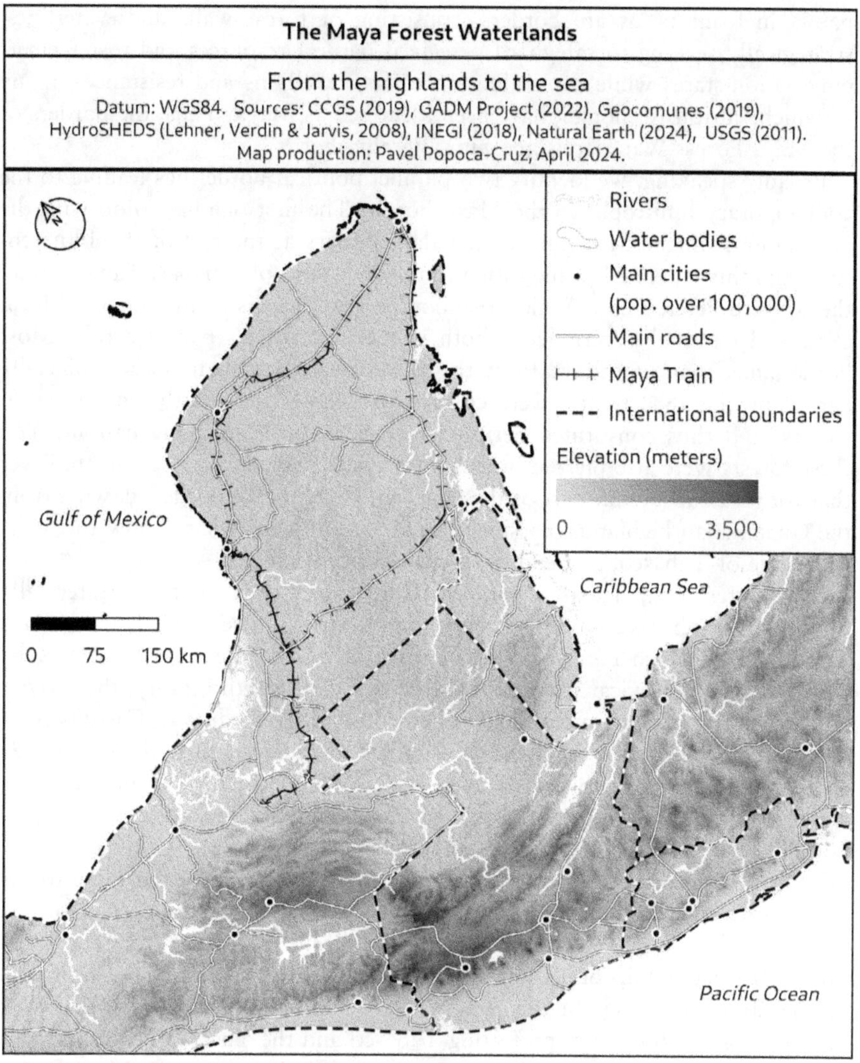

Map 0.2 Geopolitics of the Maya Forest Waterlands: From the Highlands to the Sea
Source: Pavel Popoca Cruz

We start the walk from the western corner of the Mexican Maya Forest: the states of Chiapas and Tabasco and the formation of their borders and main protected areas. We then continue along the Mexican borderlines toward the states of Campeche, Yucatán, and Quintana Roo. We then cross to Belize and finally to the department of Petén in Guatemala, ending in the place from where we set out: the western borders. To capture the mapping of the contemporary borders of the Maya Forest Waterlands, we ask the reader to bear with us as we need to jump back and forth in time. Yet, what this exercise importantly paints in front of us are borders, consisting of forest walls and waterlines, strategically seeking to safeguard perceived natural resources and foster emerging nation-states while controlling uprisings, rebellions and resistances, many of which, curiously enough, have been identified as Maya. Thus, the borders of the Maya Forest Waterlands are quite literally that.

Broadly speaking, we identify two parallel political borderings leading to the contemporary limitrophe in the Maya Forest. The first one has to do with the negotiations on the Mexican–Guatemalan borders at the end of the 19th century, and this is related to the formation of the states of Chiapas, Tabasco, and the western Petén areas. While these border negotiations mainly involved local elites and national governments, both waters and forests played a role. Most borderlands that nowadays form the Mexican–Guatemalan border, officially agreed by the 1882 treaty, were covered in deep – and mostly unexplored – forests, and thus constituted peripheral areas of both emerging nations. Yet, these forests were also pierced by powerful rivers, such as the Usumacinta River that forms a substantial part of the Mexican-Guatemalan border, flowing from the Guatemalan highlands, passing Chiapas and toward what is now the Mexican State of Tabasco, and ending in the Gulf of Mexico.

This latter area, Tabasco and the Gulf, has always been a strategically important water access and resource for Mexico. These same lands form a narrow strip, which represents the Mexican access to the Yucatán Peninsula. Thus, from the Mexican viewpoint at the end of the 19th century, these riverlands of access and resources had to be protected. According to Castillo et al. (2006), the vast southern highland forest areas surrounding the riverlands served as a "strategic wall", which had to be included in Mexico to ensure access to water resources and to the eastern lowlands. The main cities near the planned border were Comitán (Chiapas) and Tenosique (Tabasco), with particularly difficult access to the vast rainforest corridor that extended from these towns toward the largely uninhabited Petén. However, for this same reason, on the Mexican side, Chiapas was considered a strategically important "forest wall" to protect the waterlands of Tabasco, which also stood for access to the Gulf, and the Yucatán Peninsula. In this way, the formation of Chiapas with its vast rainforest was more about protecting Tabasco and the Yucatán Peninsula.

The second important creation of geopolitical limitrophe in the Maya Forest concerns the "heart of the Maya Forest" (Ankersen & Arriola, 2001), that is, the broad area of northeast Petén, Belize, and the Yucatán Peninsula. These borderlands are what Castillo et al. (2006) call the "Maya border" (p. 85), referring to the vast, thick lowland rainforest that served as a Mayan fortress and allowed the building of an autonomous Mayan force prior to the Caste War in Yucatán, discussed below. While the strongest Mayan polity was located in what is now the state of Quintana Roo, prior to the contemporary borders, no clear borders existed between Petén, Belize, and Yucatán. Rather, these were created as a result of the Mayan rebellion, although vast areas of Petén were relatively unaffected by the war. Yet, the Mexican southern states of Campeche, Yucatán, and Quintana Roo were established to contain the

Mayans and their natural resources, and the same events led to the negotiations over the Mexican–Belizean–Guatemalan borders in this area, while stirring human mobilities across the region.

Map 0.2 shows the international borders from a different, geopolitically strategic angle pictured from the highlands toward the sea. As can be perceived, the area that is Chiapas in Mexico today, which back in time was covered with deep forest, provided what may look like a defensive wall to protect what is now the narrow strip of Tabasco's waterlands and the Mexican access to both the Gulf and the Yucatán Peninsula. In contrast, the areas now comprising the three other states of Mexico – Campeche, Quintana Roo, and Yucatán – seem remote from the political center of Mexico and closely connected to Belize and eastern parts of the Petén, with access to the Caribbean Sea.

Our trail starts in the western area of the Maya Forest with the state of Chiapas, which has been considered a rugged terrain that was isolated from colonial authorities. Aside from the Mexican revolutionary battles, it has a long history of persistent Indigenous revolts (e.g., Viquiera & Ruz, 2004). Chiapas' leading elites struggled to form an autonomous state, although it was eventually attached to Mexico in 1824. Given this development, the *finca* system, which refers to estates or ranches that worked as plantations and were based on peasant exploitation, endured in Chiapas much longer than in many other places, eroding in waves from the beginning of the twentieth century onward. The peasants, whether to escape the *fincas* or the eroding system, moved mainly toward the Lacandon rainforest, where they hoped to find free land.

The history of peasant exploitation and continuing land ownership issues later gave rise to increasing revolts in Chiapas, first organized around many peasant organizations in the 1980s and then culminating in the armed Zapatista uprising in 1994. With the Zapatista autonomous territory in Chiapas, many new borderings and territorializations took place – among others, what has been considered a counter-insurgency tactic of remunicipalization. In 1999, the strategy divided the southern parts of the great Ocosingo municipality, which included most of the Lacandon rainforest and the "conflict zone", into the new, smaller municipalities of Marqués de Comillas, Maravilla Tenejapa, and Benémerito de las Americas, in addition to four other new municipalities within the state (e.g., Leyva & Burguete, 2007). The objective of this new political bordering of the vast Lacandon rainforest area was to contain and control the rebel territories, although in some cases, it also corresponded with the claim of the local population, such as in Marqués de Comillas. The Montes Azules Biosphere Reserve was one of the first to be created in Mexico in 1978. It now covers and protects most of the remaining rainforest and is surrounded by other, smaller protected areas, as shown in Map 0.1.

It is noteworthy that in Chiapas, as in all southern Mexico, nearly half of the lands remain socially owned, pertaining to communal and *ejido* ownership created after the 1910 Mexican Revolution (e.g., Martínez et al., 2023). The Mexican social landownership system is one of a kind – a rarity at the global scale – and has important implications as to how conservation and development projects can be employed. In tropical southern Mexico, most of the protected areas are categorized as biosphere reserves precisely because of the system of social landownership, which means that communities and their lands have to be involved in conservation plans, if they are meant to succeed at all. Thus, conservation in Mexico could never be about empty wildernesses composed solely of uninhabited national parks. Yet, this land ownership system has had its peculiarities in the case of the Montes Azules Biosphere Reserve, as will be discussed further along in this introduction.

The state of Tabasco, neighboring Chiapas, was also founded in 1824. As already suggested above, Tabasco can be described as the wetlands of the Maya Forest, as it is located in the borderlands between lands and waters in many ways: along and in-between the Usumacinta and Grijalva Rivers, which divide the state in two different regions. Originally, these rivers served as the Mayan routes of connectivity, and they end in the wooded coastal wetlands currently identified as an ecoregion and designated as the Pantanos de Centla Biosphere Reserve in the Gulf of Mexico.

Tabasco also holds a long history of Indigenous revolts, in addition to French and US invasions, and conflicts with the state of Yucatán. Strategically, Tabasco contains a third of the water resources of Mexico and considerable oil reserves. In the 1950s, and with the ending of the principal trade related to precious woods exploited along the Usumacinta River, Tabasco was to be transformed into a modern agricultural paradise, deforested and eroded by vast cattle-ranching areas (Laako & Kauffer, 2021; 2022). Given the extensive deforestation, Tabasco has occasionally been excluded from the Maya Forest area definitions (see discussion in Laako et al., 2022). Yet, these accounts have failed to recognize the key role of waterways for such a space as the Maya Forest, because both the Mayas and the forest depended on their waters. Thus, in the predominant definitions of the Maya Forest, waters seem to have been hidden and separated from forests and lands, as shown by the exclusion of Tabasco. Tabasco may be considered as the borderlands of the Maya Forest: seemingly at its edges, but also at the crossroads of several waterways toward the Yucatán Peninsula and the Gulf of Mexico, and with its rivers connecting to the Chiapanecan and Guatemalan highlands and forests.

Continuing from the strategic waterlands of Tabasco toward the Yucatán Peninsula, we enter a world that has also been called the *Mayab*. Indeed, the contemporary political borderings increasingly relate to the history and control of the Mayas. Previously a vast Mayan territory, this area was also the stage for the Caste War of Yucatán (1847–1915), fought between the native Mayas and *Yucatecos*, the Hispanic population, which initially resulted in a Mayan victory and the creation of the state of San Chanta Cruz between 1847 and 1883 (e.g., Castillo et al., 2006). Eventually, however, the war ended with the Mexican army occupation in 1901. The Caste War had its roots in the Mayan peasantry revolt against the loss of land ownership and harsh working conditions in the *henequen* haciendas, which had intensified in late colonial and post-independence Yucatán and had brought about new white and mestizo migration to rural towns, offering hope of economic opportunities. The Mayan state of San Chanta Cruz initially received support from the United Kingdom, because they shared profitable trade relations in the Belizean territory. The Mexican victory was not achieved until Mexico negotiated with Britain the Belizean border (back then, British Honduras). These Mexican–Belizean negotiations ended diplomatic relations with the Mayan State and defined a new international border between Mexico and British Honduras finalized in 1897 (e.g., Castillo et al., 2006; Toussaint, 2009; Shoman, 2020).

The Caste War also affected the interior state-borderings within the Mexican Yucatán Peninsula. The state of Yucatán was founded in 1823, and in 1841 it declared independence from Mexico as the Republic of Yucatán. It rejoined Mexico again in 1848, and was first divided in the midst of the Caste War with the creation of the state of Campeche. Officially, Campeche was founded in 1863, and it holds an important shoreline with access to the Gulf of Mexico and oil reserves. It also contains large rainforest areas bordering Guatemala and Tabasco with some important rivers and small lakes, such as Palizada, Candelaria, Champotón, and the Términos Lagoon. The city of Campeche was declared a World Heritage City in 1997, and the state has three biosphere reserves: Calakmul (1989), Ría Celestum, which is also a Ramsar wetland (2004), and Los Petenes, also a Ramsar wetland (2004).

The state of Quintana Roo is the final rebordering rooted in the Caste War and the very location of the Mayan stronghold: the independent Mayan state of Chan Santa Cruz, now the town of Felipe Carrillo Puerto, bordered by the Sian Ka'an Biosphere Reserve, was created in 1986 and also declared a UNESCO Heritage Site in 1987. Formally, the state of Quintana Roo was established in 1974; however, prior to this, it had also been defined as the Territory of Quintana Roo and, for a short time, formed part of the state of Yucatán. There have been some negotiations on the border between Quintana Roo and Campeche. Originally, Quintana Roo shared a short border with Petén, Guatemala, but currently this area pertains to Campeche.

Nevertheless, the state of Quintana Roo, a coastal area, characterized by marine ecosystems as well as tropical forests, shares a long history of Mayan struggles and relations with neighboring Belize (e.g., Pozo et al., 2011). Nowadays, most of these coastal areas are protected: the Chetumal Bay has been designated as the Manatí wildlife conservation area since 1996. The coastal area of the southeastern tip of Quintana Roo, consisting of coral islets, is the Banco Chincorro Biosphere Reserve, created in 2006. The Biosphere Reserve of Caribe Mexicano Profundo is also subject to the federal governmental agreement from 2018, but is currently not found on the UNESCO list of Mexican biosphere reserves (SEGOB, 2018).

In this way, the *Mayab* – currently the state of Yucatán – is a reduced version of its original territory. It is also the stage for the *henequen auge* (approx. 1890–1910) and the Caste War. Owing to its seashore, it has always held global connections as part of the shared Atlantic borderlands. As we are writing these lines, Yucatán has only state-level and municipal protected areas – in other words, no federal ones. Yet, these importantly include, for example, the Ring of Cenotes (*Anillo de Cenotes*), designated as a Ramsar site in 2009, which comprises a complex of 99 *cenotes* or sinkholes and is identified as a unique water system in Mexico and globally, being the product of a large meteor impact 65 million years ago (Ramsar, 2009). These can be consulted in Map 0.1. The *cenotes* also serve as resting grounds for waterfowl during their migration to the South and hold endemic species of reptiles and others. Worth mentioning, but so far with little information available, is the Puuc Biocultural State Reserve, created in 2011 and consisting of tropical, dry forest in the Central Yucatán Peninsula. It seems to be the first of its kind in

Mexico, with its aim to preserve the biological diversity of the region, including its historical value for the Mayan culture and sustainable livelihoods (Portillo-Quintero et al., 2023). It covers the territory of five municipalities traditionally dedicated to *milpa*, communal forestry, and cattle ranching in the form of communal and private properties, and has been created as a corridor connecting other protected areas toward Campeche and Quintana Roo.

Belize, the former British Honduras, has historically been a country of forestry rather than agriculture, and for this reason, it has a different political and economic history than Guatemala and Mexico, although culturally, they share the same roots as indicated above. Currently, Belize has no biosphere reserves, although some were actively planned in the 1980s and 1990s, for example in the areas which are today the Mountain Pine Ridge Forest Reserve, Chiquibul National Park, and, particularly, the Caracol Archaeological Reserve. It was also considered to propose the latter as a World Heritage Site. The interviews with conservationists in Belize by Laako in 2023 indicated that the lack of biosphere reserves was due to the fact that Belize does not have a settled, permanent population within the planned area, while the biosphere reserve program is particularly aimed at sustainable development. Apparently, there was also a similar idea to create a biosphere reserve in southern Belize in the area of the Sarstoon-Temash National Park, but it was abandoned owing to a lack of funds (Sarstoon-Temash National Park – Transcript of Stakeholders' Workshop, 1997). However, Belize does have a UNESCO Natural World Heritage Site called the Belize Barrier Reef, created in 1996, and the Sarstoon Temash National Park was designated a Ramsar-site in 2005.

Belize is a small (less than half a million citizens) but extraordinarily diverse country composed of inhabitants of Mayan heritage, *creoles*, *mestizos*, Garifuna people, Mennonites and people of mixed European, Central American, Chinese and Caribbean ancestry, among others. Among the six existing districts, the Toledo District in southern Belize is the location of the customary Mayan lands, although this is not to say that no Mayas can be found elsewhere in Belize. For example, the western San Ignacio area also has strong Mayan heritage.

As has been the case with many other national or international borderings in the Maya Forest, the Mayas are also implicated in the case of Belize. At the same time as the Belizean Mayas are struggling for their land rights despite a consent order issued by the Caribbean Court of Justice in 2015, the state of Belize is also involved in a case at the International Court of Justice concerning the contested border with Guatemala, dating back to colonial times. Thus, two bordering struggles are taking place simultaneously, centering on – fully or partially – the same Indigenous lands: the area of the international border between Guatemala and Belize and that of the Belizean Mayas within the Belizean State. The two are connected not just geographically, but also owing to the arguments of the Belizean state, which denies the Mayan customary rights on the basis that the appellants were Q'eqchi and Mopan Mayan immigrants from Guatemala, who would not satisfy the test of pre-sovereignty occupation and continuity as set in Commonwealth decisions. Yet, during the colonial era, Indian reservations were established along the border to show that these communities belonged to the empire (Wainright, 2008).

Guatemala and Britain reached an agreement on the disputed border area in the Wyke-Aycinena Treaty in 1859. The treaty recognized British sovereignty over the contemporary Belizean territory back then pertaining to Guatemala, later leading to the creation of British Honduras. Prior to the Wyke-Aycinena Treaty, Britain and Spain had settled on various treaties establishing that the Belizean territory pertained to Spanish sovereignty, though British settlers could use the land. Despite the Wyke-Aycinena Treaty, Guatemala has often laid claim to Belizean territory and questioned the Wyke-Aycinena Treaty, with this issue now, in 2024, to be resolved by the International Court of Justice. The Guatemalan claims have occasionally referred to the southern parts of Belize and the whole Belizean territory (e.g., Shoman, 2020).

These border conflicts are also present in the fluid edges between Belize, Honduras, and Guatemala around the Sarstoon River. For Guatemala, access to the sea is an important claim, and as a result, over the years, the treaties between the countries have specified unrestricted Guatemalan access within two miles around an equidistance line that divides the Belizean and Honduran sea territories. Additionally, an attempt was made to establish a trinational ecological park in this border area (e.g., Shoman, 2020; Toussaint, 2009). However, as noted by Toussaint (2009) and Shoman (2020), the borderlines drawn in both the Hondo River between Mexico and Belize and the Sarstoon River between Guatemala and Belize have caused disagreements owing to previous unclear and erroneous mapping, which did not show the border in the rivers where it was supposed to be. This is a problem of ancient mapping, border delineation, and seasonal variations of the natural flow but, above all, an issue of the relationships between states. Borders change when the natural elements defining them change; in other words, fluid edges – unstable, capricious, erratic, intermittent and also vague – are concepts constructed by humans.

According to Wainwright (2008), the unsettled nature of southern Belize itself is due to the fact that the treaties between the colonial powers of Spain and England only covered the land until the Sibun River, located near the center of the country. Despite suspected contact early on, the southern part remained uncolonized, with the exception of a few loggers exploring the territory, and contested at least until 1880 (Wainwright, 2008). However, by the 1880s, thousands of Q'eqchi Mayas had fled Verapazes to the north, Petén, and to the lands along the rivers in the east. Wainwright (2008) notes that the existing Mayan communities composed of Maya-speaking Q'eqchi', Mopan, and Manche-Chol, grew with the influx of the Mayas from the Toledo District – "a political space that did not yet exist". Simultaneously, the mahogany boom continued, resulting in increasing logging concessions granted in southern Belize, including to multinational companies. According to Wainwright (2008), by 1787, 12 settlers owned most of the land in British Honduras. However, with the short-lived boom soon ending in bankruptcy, the lands became the property of the colonial state, which sold them on as private property. Given the timber extraction and land monopolies, agriculture lagged behind forestry practices, and no land reform took place. Thus, most Mayas still lack land and tenure.

The contemporary struggle of the Belizean Mayas is related to logging concessions granted in the 1990s, which the Mayan communities first challenged in the Supreme Court of Justice of Belize, followed by the Caribbean Court of Justice, which ruled in their favor in 2015, although the Belizean government has so far failed to implement the Court's decision. Thus, contested Mayan and geopolitical boundaries persist in the Belizean–Guatemalan borderlands.

As landownership in Belize is dominated by private property, in recent years, nature conservationists have adopted a new strategy and now purchase private land for protected areas. For example, a considerable effort has been made in Belize to create conservation connectivity with the Maya Forest, particularly by the non-profit Maya Forest Trust Fund. This comprises more than a dozen conservation organizations that, in 2021, purchased an area of 96,000 hectares located on the Belizean–Guatemalan border area next to the Río Bravo Management and Conservation Area, managed by Programme for Belize, a local NGO, and bordering the trinational corner of Belize, Mexico, and Guatemala (e.g., Global Conservation, 2024; Laako's interviews in Belize in June 2023). The area protected by the Maya Forest Trust Fund is not included in the governmental program of protected areas, but according to Laako's interviews with conservationists in 2023, the land is considered to be more in "public Belizean hands" than the earlier private land owned by companies. Also connected to this new protected area is the Maya Forest Corridor project coordinated by the Maya Forest Corridor Trust, which also purchased land around Central Belize, between Belize City and Belmopan and bisected by the Western Highway. This project aims to create connectivity toward the protected areas located in the northwestern corner of Belize and in the south and west, as well as to protect the fragile nature in the most densely populated area of Belize. In particular, the Maya Forest Corridor has focused on jaguar conservation.

On the other side of the contested "adjacency zone", the official term preferred by Guatemala, is the Chiquibul Maya Mountains Biosphere Reserve, created in 1995. According to the Guatemalan Council of Protected Areas CONAP (2019) applying to the area, the complex has great "strategic value" for its unique ecosystem, presence of cultural heritage, and geopolitical importance, including its connectivity to the Maya Biosphere Reserve located in the northern part of Petén and the "Maya Forest" as well as the "adjacency zone with Belize" with pending territorial and maritime issues to be resolved. Formally managed by the Guatemalan CONAP together with the Guatemalan Institute of Archaeology, Ethnology and History (IDAEH) in collaboration with Asociación Balam, nearly 80% of the forest coverage have been lost. The area has been populated by the Mopan and Manche-Chol Mayas and later occupied by the Q'eqchi. There is considerable tension along the border, not only owing to the current claim taken by both states to the International Court of Justice in 2019, but also because of pressures from the Guatemalan side on the Belizean Chiquibul National Park in form of illegal logging, cattle ranching, and biodiversity smuggling, which have caused incidents between the Belizean armed forces and trespassing Guatemalans. This is why the Guatemalan Asociación Balam and the Belizean Friends for Conservation and Development (FCD) have developed a considerable strategic and joint binational effort toward biocultural conservation, which would allow to halt deforestation, illegal logging, and smuggling and to improve the livelihoods of the communities in the Chiquibul Maya Mountains Biosphere Reserve.

The department of Petén, created in 1866 and covering the northern part of Guatemala, was relatively unoccupied until the 1960s, when the Guatemalan government began to promote the colonization of the forested woodlands along the Usumacinta River and its tributaries. The creation of the national company *Fomento y Desarrollo de Petén* (FYDEP), later transferred to the National Agrarian Institute (INTA), converted the colonization process into a state's policy – and contributed to the further militarization – of the Petén in the context of the internal armed conflict. As reported by many authors (e.g., Ybarra, 2018), for the Mayas, the effects of the Guatemalan Civil War have been profound in terms of migration, displacement, massacres, and loss of landownership. It created a massive Mayan diaspora throughout Canada, the United States, Mexico, and Central America (e.g., Loucky & Moors, 2000) mainly fleeing from the southern departments of Huehuetenango and El Quiché.

The Maya Biosphere Reserve, covering nearly half of Petén, was created in 1990, when the Peace Accords were still to come. As a result, conflicts have arisen between inhabitants and conservationists. Ybarra (2018) reported the problematic and often violent dynamics between land ownership, livelihoods, and conservation in the case of the Laguna Lachúa National Park (located in the southern Huehuetenango Department) and in some parts of the Maya Biosphere Reserve within Petén. The Maya Biosphere Reserve includes many other protected areas and the multiple-use zone, which consists of communal forestry concessions, managed by the Association of Forestry Communities of Petén (ACOFOP). The ACOFOP, created in 1995, represents 24 communitarian organizations in Petén. It maintains that the Maya Biosphere Reserve was created precisely because the border area was affected by the internal armed conflict (Laako's interview in Flores, June 2023).

The Petén Department comprises the largest concentration of archaeological sites discovered in the Maya Forest. Petén is a historical borderland of Indigenous people, crossed local and current international migrations, and state colonization-induced projects as well as oil explorations and extraction in the Laguna del Tigre National Park, which have strongly contributed to deforestation. Petén, the largest department of Guatemala, has historically been a territory of extraction – *chicle*, timber, petroleum, livestock, palm oil – and illustrates the great contradictions between state policies of colonization, the aftermath of the armed conflict, and the conservation policies from the 1990s. More recently, land grabbing and land accumulation by drugs traffickers have been successful logistics strategies, especially in remote locations of the Maya Forest Waterlands, mainly through the conversion of forests into pasture – and more recently, monoculture such as palm oil – and owing to money laundering facilities (McSweeney et al., 2017).

Finally, our trail throughout the Maya Forest geopolitical borderings ends again at the Usumacinta River in the Guatemalan–Mexican borderlands. In this Mexican–Guatemalan corner along the Usumacinta, three protected areas are in close proximity: the Montes Azules Biosphere Reserve in Chiapas in the south, the Usumacinta Canyon Wildlife Refuge in Tabasco in the north, and the Sierra del Lacandón National Park on the Guatemalan side.

Environmental historians (e.g., Miller, 2007) have long shown that the region now called the Maya Forest was "rewilded" during colonization. With the dramatic decline of the pre-contact population from about five million to one million, most areas were regrown with secondary forests now subject to conservation. The Maya Forest landscapes continue to appear mostly rural, with an estimated 600,000 rural inhabitants according to the previously mentioned *Selva Maya Programme*'s website. Yet, population numbers are increasing as a result of several governmental colonizing and developmental programs employed since the 1950s. As shown in Map 0.2, among several expanding cities, Mérida and Cancún both account for over 800,000 inhabitants, now connected by the Maya Train.

This is the brief geopolitical trail of the region, mapped and bordered by the given countries and their inner and cross-border organization and mobilities, which involve complicated bio- and geopolitical dynamics. Next, we turn to explain and outline the three transboundary and connecting corridors we have identified in the Maya Forest waterlands, which also provide the structure for the rest of the book.

Mapping the Transboundary Corridors: Lacandon Forest Waterlands, Laguna del Tigre-Gulf Riverlands, and Caribbean Karst Waterlands

We have now taken the reader to a trail mapping the main geopolitics in the Maya Forest, highlighting two historical borderings in particular: the first one related to the western highlands covered with forest and ending in riverlands flowing toward the Gulf of Mexico, and the second to the Yucatán Peninsula, which included Belize and the eastern parts of Petén looking toward the Caribbean Sea. Although the above section focused on the formation of national borders and those of the main protected areas, a careful reader may have detected by now certain corridors within the Maya Forest that share some unique historical, social, and natural characteristics despite being divided by political borders. These corridors form units that help to structure and examine the broader tendencies of the Maya Forest.

We have thus identified three transboundary corridors within the Maya Forest: the Lacandon Forest Waterlands, the Caribbean Karst Waterlands, and the Laguna del Tigre-Gulf Riverlands. These corridors illustrate the nexus of Maya/forest/water/lands in distinct and shared ways. They are not fixed categories, nor do they have precise borders; rather, they are blurred, vague, and fluid, and overlap with each other (see Map 0.3). Yet, they allow exploring certain developments in the Maya Forest that are differentiated, connected, and

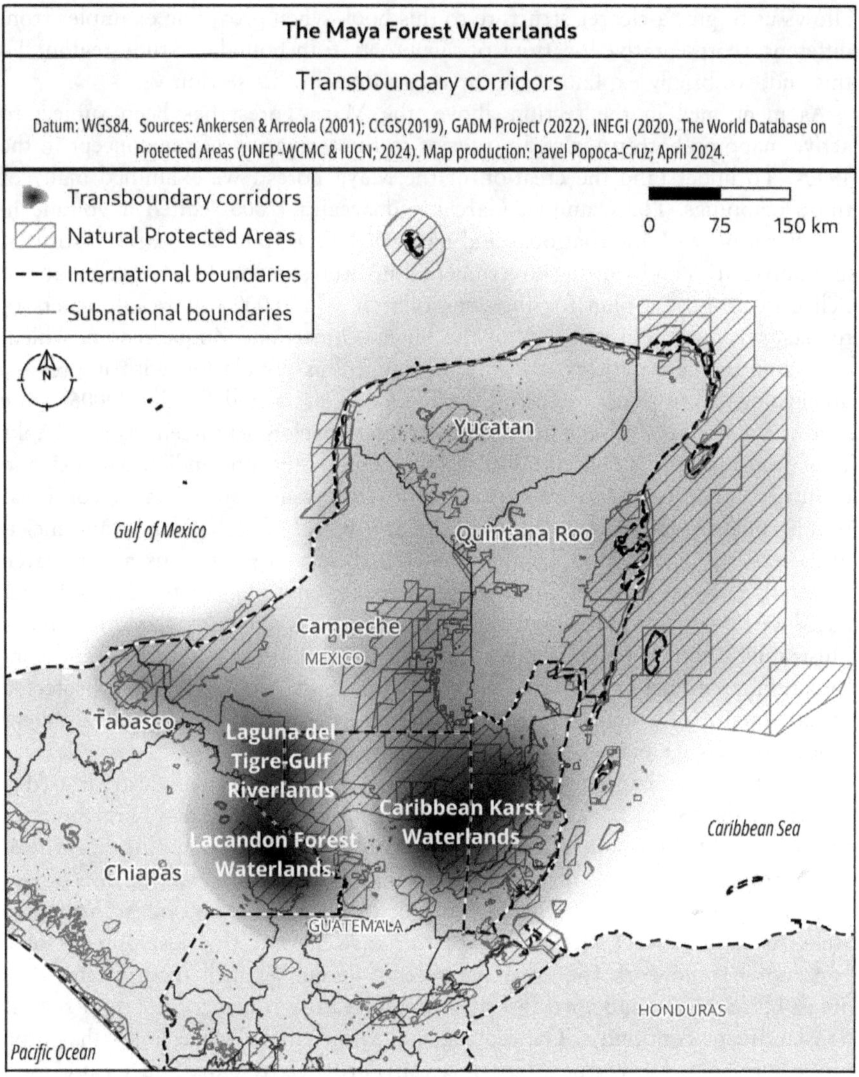

Map 0.3 The Three Corridors of the Maya Forest Waterlands
Source: Pavel Popoca Cruz

entangled. We have used these corridors to examine the ways in which certain elements (forests, waters, lands) and people relate to the pre-established categories and existing borderings, which are at the very heart of this book.

Thus, we have – inevitably perhaps – engaged in our own mapping, based on our experience in the field across the territory of the Maya Forest Waterlands and on the literature reviews and analysis conducted for this book. At the same time, this mapping, which surfaced from our work somewhat unexpectedly, allows us to give a clearer structure to this book when providing examples from different, representative locations of our broad, transboundary study region. To this end, we briefly explain these three corridors in this section.

As mentioned in the section above, the Maya Forest has been subject to active mapping both prior and posterior to the formation of the concept in the 1990s. To understand the creation of the Maya Forest, we examined many of those mappings. For example, García and Secaira (2006) edited a volume in collaboration with international and national NGOs, public research institutes, and universities and various governmental institutes in Mexico, Guatemala, and Belize to propose a plan for the conservation of 190,000 square kilometers of forests, savannas, and wetlands of the Maya, Olmec, and Zoque region, which, according to these authors, harbors an Indigenous population with impressive archaeological vestiges. In their introduction, García and Secaira (2006) state that as the national governments and different development agencies and banks were focusing on the conservation of large regions (beyond individual sites and countries), "Mesoamerica must generate a wider vision and broader conservation agenda for adequate insertion in the new world of globalization and mega-initiatives" (p. 5). To do this, the nine civil society organizations and research centers had come together to develop an ecoregional plan of the Maya Forest based on a network of protected areas. The book contains numerous maps illustrating different issues, from social and natural aspects to threats and conservation areas networks. The latter ones include, among others, protected areas, aquatic and fluvial systems, areas rich in endemic species, and different models of management.

Prior to these mappings, Ankersen and Arriola (2001) had also drafted a report for the "Selva Maya Coalition and the Tropical Ecosystem Directorate of the UNESCO Man and Biosphere Programme". This mapping identified, among others, the work of conservation biologists, who, since the 1960s, had sought answers to address the problem of forest fragmentation. According to Ankersen and Arriola (2001), one of the answers was the development of a conceptual framework for "regional reserve design", which incorporates core areas, buffer zones, and corridors allowing migratory routes for different species and landscape continuity. The eco-regionalization of the Maya Forest has been an answer to conserve the Mesoamerican Forest ecosystems and ensure their (transboundary) continuity.

We paid particular attention to the work of Ankersen and Arriola (2001) as their identification of three (ecological, conservation, and development) corridors of the Maya Forest evolved around different socionatural characteristics and initially coincided with our analysis of historical borders and political processes. The three main corridors of the Maya Forest according to Ankersen and Arriola (2001) are (1) The Lacandon Corridor, comprising the Lacandon rainforest and the Sierra del Lacandón intercepted by the Usumacinta River Basin (Mexico-Guatemala); (2) the Ceibo Corridor comprising the Sierra del Lacandón and Laguna del Tigre National Parks, intercepted by the San Pedro River and extending to the Maya Biosphere Reserve (mainly Guatemala); and

(3) the Belize Corridor and the "Heart of the Forest", which covers the corner between the Calakmul Biosphere Reserve, northwestern parts of the Maya Biosphere Reserve, southern parts of Quintana Roo, and the whole Belizean-Guatemalan borderlands area (mainly Belize and Guatemala).

We want to emphasize that our notion of a corridor is distinct from that of Ankersen and Arriola (2001) and the others mentioned above. We are not seeking developmental corridors or ecological/biological corridors ripe for further conservation. Our corridors are identified on the basis of socio-environmental features, based on some shared social, historical, and political processes and environmental aspects. However, as the work of Ankersen and Arriola (2001) is also dealing with some of these aspects, we used it as a basis. These aspects are explained in more detail below (see Map 0.3).

In terms of the first corridor, the Lacandon Corridor, Ankersen and Arriola (2001) identified the broader block of the rainforest, empowered by the Usumacinta River, which forms the geopolitical border between Guatemala and Mexico, but also embodies the shared watershed area that includes most of this transboundary region. In fact, in our consideration, the western fringes of the Maya Forest are limited to the Usumacinta River basin in Chiapas, rather than covering the whole state.

According to Ankersen and Arriola (2001), the Lacandon Corridor also includes an area with a particular land-use history and a unique political unit under Mexican law, which granted the Lacandon Community, a forest-dwelling Indigenous group on the Mexican side, 614,000 hectares of forest lands to practice their traditional systems of agroforestry in the 1970s. This process generated deep conflicts and evictions of other groups within the rainforest, some of whom were Zapatistas. Both the Zapatistas and the Guatemalan Civil War, with its Northern Transversal Strip (NTS) on the other side of the Mexican southern land border, brought about strong militarization and Guatemalan refugees on the Mexican side of the border.

We agree that within the Maya Forest, the Lacandon Forest Waterlands (see Map 0.3) represent this sort of shared corridor, which we identify as transboundary, characterized by the forests of the Guatemalan highlands and the continuing rainforests of Chiapas, shared by the Usumacinta River and its transboundary tributaries (see also Carabias et al., 2015). Socially and historically, the inhabitants have shared the space, which has been fairly remote from the other urban or political centers. Our own fieldwork has been predominantly focused on this borderlands area, where we have witnessed inhabitants moving and communicating across the borders. Moreover, the name, "Lacandon", appears on both sides of the border, as in Lacandon rainforest in Mexico, and the Sierra del Lacandón National Park in Guatemala.

The second one, the Ceibo Corridor, was identified by Ankersen and Arriola (2001) as a wide belt of fragmented landscape that connects to Guatemala's main protected areas, the Laguna del Tigre National Park and Maya Biosphere Reserve, including the transboundary San Pedro River, which, on the Mexican side, reveals the deforested cattle pasture of the Tabasco lands. The Ceibo, which is the name for the border crossing area between Guatemalan Petén and Mexican Tabasco represented the main colonization front of strategic importance in terms of transportation and commerce, leading both to the oil fields and to the main population center, El Naranjo, which was first created as a *chicle* and logging camp on the banks of the San Pedro River in Guatemala. These were also the lands of illicit trade of artifacts, wildlife, timber, and human cargo, or what the authors call Guatemala's "wild, wild West".

While we mostly agree with the above identification, we ended up with a slightly different – and more transboundary and waterier – mapping of this corridor. Based on our fieldwork in many locations of Tabasco, and especially along the Tabasco–Guatemalan borderlands, the San Pedro River on both sides of the border, and including the eroded wetlands in the Laguna del Tigre National Park, now controlled by drug cartels, we decided to name this corridor the Laguna del Tigre-Gulf Riverlands (see Map 0.3).

In our mapping, the heart of the corridor includes the Ceibo, but also the point of coming-together of the rivers, where the highlands end and the low riverlands continue to the Gulf of Mexico as the historically strategic resource and access place. This is where, historically, the rivers formed an access, also called the "Mayan rivers", toward the deeper rainforest with its ruins and resources; this is where these resources were taken to to be transported to the port. Thus, the corridor importantly includes the meeting place of Canyon of Usumacinta, connecting Guatemala, Tabasco, and Chiapas, as well as the city of Tenosique, where many timber sources were taken to on their commercial route. In a similar vein, as the corridor includes the oil reserves of the Maya Forest, it also extends toward the Campeche coastal areas. Thus, the Laguna del Tigre-Gulf Riverlands begin at the wetlands of the Laguna del Tigre National Park in Guatemala, grow strong in the axis of the Ceibo-Canyon of Usumacinta-Tenosique, where they also overlap with the Lacandon Forest Waterland Corridor, and then flow with the rivers toward the coastal lowlands of Tabasco and Campeche – the Gulf of Mexico.

In this way, we seek to emphasize the waterlands of the Maya Forest with its historical and strategic characteristics – which also include the Maya – and its contemporary phenomena.

Finally, in terms of the Belize Corridor and the "Heart of the Forest", Ankersen and Arriola (2001) identified the northeastern parts of the Guatemalan Maya Biosphere Reserve, connected to the Mexican Calakmul and to the rainforests on the Belizean side, particularly the Río Bravo Conservation and Management Area and the Chiquibul National Park and Bladen. The authors described this corridor as an ancient and contemporary trade route, which connects the Mayan ruins and archaeological sites with today's tourism, centered around the development of the Maya Route within the Maya World and including, for example, Tikal (Guatemala), Calakmul (Mexico), and Caracol (Belize). This area also features the special system of forest concessions within the Maya Biosphere Reserve, designated as the multiple-use zone. The road from Flores, Guatemala, to the border with Belize and from there to Belmopan and Belize City forms an old connection path to the Caribbean Sea.

Again, while we mostly agree with this proposition of the corridor, particularly its "heart area" surrounding the eastern parts of the Maya Biosphere Reserve and Belize, we greatly extended it toward the Yucatán Peninsula, including the Mexican states of Campeche, Quintana Roo, and the Yucatán as part of the same extending lowland and coastal region deeply affected by the Mayan borderings and strongholds. We call this the Caribbean Karst Waterlands (see Map 0.3).

We emphasize how this transboundary corridor has traditionally and strategically looked toward the Caribbean Sea and also been influenced by those coastal and marine connections. Even today, as illustrated by the border conflicts, access, resources, and routes to the Caribbean Sea play an important role. In fact, the protected areas of the Maya Forest have also greatly expanded toward the coastal and sea areas (see Map 0.1), sometimes also called the Great Mesoamerican Reef Conservation Area. However, instead of separating this area into two parts (the Maya Forest and the Mesoamerican Reef), we consider it to be one entangled and interdependent system (i.e. the Maya Forest Waterlands).

While the northern parts of the Caribbean Karst Waterlands are mainly riverless lowlands of the Maya Forest, they feature the formation of the karst soil that has created a particular world of subterranean waters, *cenotes* (sinkholes), and cave systems in the higher parts, such as the Chiquibul Cave System, the largest in Belize and longest of Central America, under which the Chiquibul River flows (e.g., Mejía Ortiz et al., 2021). Thus, in addition to looking at the sea, the Caribbean Karst Waterlands also extend below the surface as a special characteristic of the region and its socioenvironmental culture.

Finally, this transboundary corridor has also been historically and socially mobile, serving as refuge in turbulent times and involving many kinds of border crossings. While pertaining to the most southern peripheries of Mexico, with strong links to Central America, it has always been connected to the outer seas and influences – both Caribbean and Atlantic – through the coastal river basin in the south.

Given these socioenvironmental characteristics, combined with certain shared political histories and geographies, the three corridors of the Lacandon Forest Waterlands, Laguna del Tigre-Gulf Riverlands, and Caribbean Karst Waterlands together form the transboundary cornerstones of the Maya Forest waterlands. In the following chapters, in which we examine different entanglements of the Maya/forest/water/lands and borderlands, we come back to these corridors and explore different issues emerging from within and between them.

Next, we turn to mapping the scientific spaces and places of the Maya Forest waterlands.

Mapping Scientific Spaces and Places in the Maya Forest Waterlands

Carmelita is a small village deep in the Guatemalan Maya Biosphere Reserve. Established in 1925, it is an old *chicle* camp, dedicated to the extraction of chewing gum, the main economic activity of the time, which tended to be located near the Mayan ruins. The town of Carmelita is also surrounded by Mayan ruins, the most important one being the ancient city of *Mirador*, which is now part of the Mirador-Río Azul National Park created in 1990. According to the Wildlife Conservation Society's website (https://guatemala.wcs.org/en-us/Wild-Places/NP-Mirador-Rio-Azul.aspx), the park and Biotopo Naahtun-Dos Lagunas are located in a borderlands area between the tall, wet tropical forests of Petén and the low, subtropical and xeric forests of Yucatán. A visit to the Mirador requires a multi-day hike from Carmelita, which currently holds a semi-monopoly over the hiking tourism to this ancient city. This is the result of a negotiation between Carmelita, conservationists, and the Guatemalan CONAP (Rahder 2020). Carmelita itself forms part of the multiple-use zone and has held a 25-year concession since 1997, organized around its own cooperative.

As can be seen in Map 0.4, these northern parts of the Guatemalan-Belizean borderlands, where Carmelita is located, are the area of the Maya Forest with the most Mayan ruins and also contain the region's thickest forest coverage. Other condensed areas of Mayan ruins are located on the Yucatán Peninsula and are dispersed over many places, but many of them follow the rivers, coastal areas, and sinkholes, *cenotes*. A substantial number of ruins are also found in the highlands of Guatemala close to the Mexican border, followed by many more in the Chiapas and Tabasco riverlands.

Carmelita and the Mirador Río Azul National Park are prime examples of contemporary Maya Forest shapeshifting realities and a showcase for the ways in which scientific research and scholars play a role in and are entangled with the nexus of Maya/forest/water/lands. As reported by Rahder (2020), a leading US archaeologist with long-standing work in Mirador has concluded that these particular ruins are, in fact, located in a broader geological basin. These findings are based on satellite data dating back 30 years and originally provided by the *National Geographic*. The pro-basin argument seems to be that the national park surrounding the ruins needs to be expanded to include and align with this supposed basin. The proposal and its active campaign have been accompanied by attempts to declare the Mirador a UNESCO heritage site. Rumors have hinted at plans for megatourist and/or private luxury tourist projects to be developed in the area. However, the proposal has also stirred strong opposition from Carmelita's cooperatives and other concessionaries in the multiple-use zone, as well as from conservationists. They oppose the plans as the new park borders would eliminate considerable parts of Carmelita's concession areas. Indeed, while the struggle seems to be focused on whether the geological basin exists or not, the main underlying controversy is between the contested borders of the Mayan ruins and community lands.

It is not our objective to dive deep into the Mirador Basin case as this has already been extensively discussed by Rahder (2020) and Devine (2016). However, we have mentioned it here as an example of the main entanglements found in the contemporary Maya Forest landscape: these, importantly, involve Mayan heritage and archaeology, the natural environment, tourism, research, conservation, borders, livelihoods, and land rights. In this section, we thus address

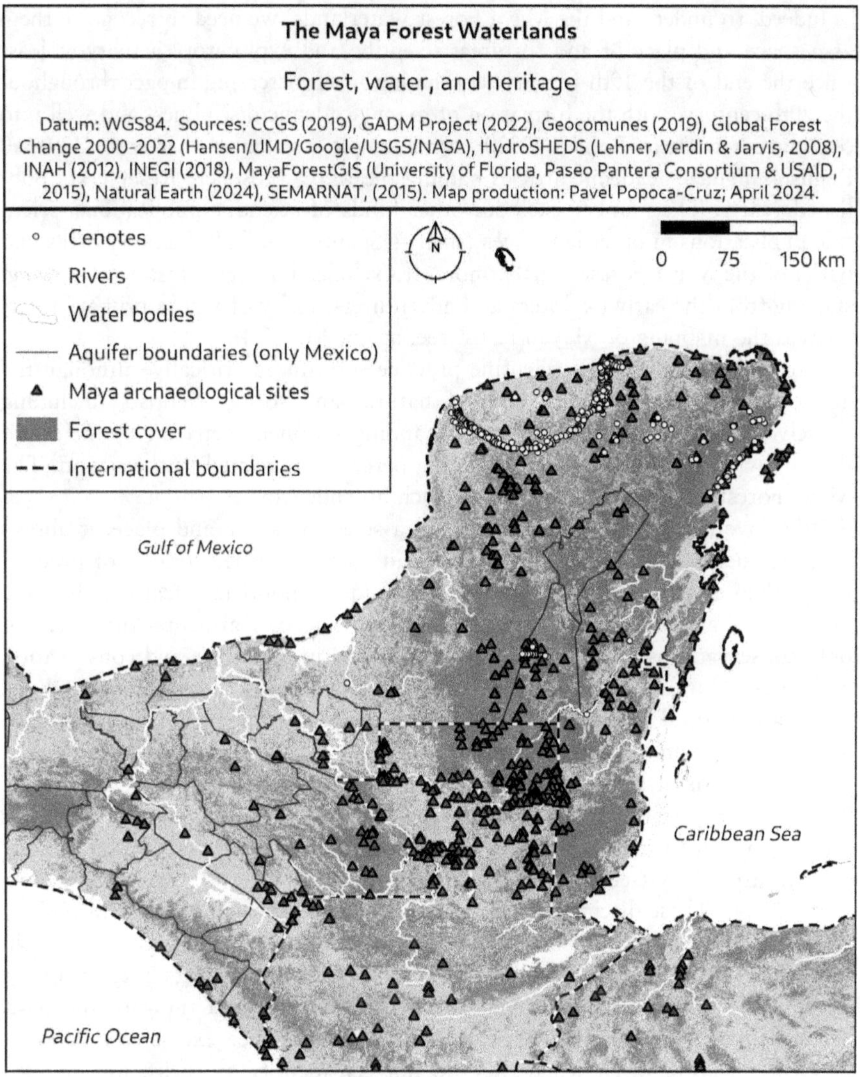

Map 0.4 Forest Coverage, Waters, and Locations of the Mayan Ruins
Source: Pavel Popoca Cruz

the other kind of Maya Forest waterlands mapping: that of scientific practice and writing, which brings together both social and natural sciences and also places them under critical scrutiny. On the one hand, these are works and perspectives on which we build as a point of departure. On the other, we critically reflect upon the scientific work both in this section and throughout this book as part of the political borderings (in which we also pertain ourselves). As the previous mappings showed, this scientific mapping also involves a space (printed on paper or digital, usually) and a place (that of the field).

Indeed, to understand the Maya Forest waterlands, we need to recognize them as a space and place of and for great scientific and exploratory activity at least since the end of the 19th century, continuing with increasing impact throughout the 20th century, with the formation of many academic disciplines, and well into our current context. Rahder (2020), for example, calculated that some 120–150 English-language academic articles naming the Maya Biosphere Reserve are published yearly. The number excludes other kinds of scientific publications, scientific publications in other languages (such as Spanish), and all the other areas and parks of the Maya Forest. Furthermore, it excludes long-term historical writing, starting from the early explorers and adventurers, many of whose primary interest was the mapping of Mayan ruins (see, again, Map 0.4).

Rarely, however, is this scientific practice scrutinized critically, although the above clearly suggests how different natural and social scientists (including ourselves) shape this region both by mapping its phenomena on the field and then by writing, publishing, and shaping perceptions related to the region. The Maya Forest, at its heart, is precisely such an endeavor.

When we address the Maya Forest as a scientific space and place, it allows locating such cases as the Carmelita-Mirador in a broader context of practice rather than as isolated cases, despite their uniquely politicized features. In these informative and formative cases, such as the one of Carmelita-Mirador, not only conservationists play a role – as implicated in ecologists and conservation biologists – but also archaeologists and anthropologists, among many others. Social and natural scientists alike, in their/our broad scope, make an appearance as powerful actors of the Maya Forest.

This analysis also invites other reflections upon academic power structures. The Maya Forest is the region that gives origin to the well-funded Mayanist Studies, which form a North American field in the interstices of Archaeology and Anthropology (see, for critical discussion, e.g., Fisher & Chase 2021; Hostettler 2004). These disciplines began to take form at the end of the 19th century in relation to the Mayan ruins, and have resulted today in heritage sites within different protected areas. This development has had many effects in terms of the Maya Forest realities. First, "Maya" often appear to be a thing of the past. Second, many contemporary people have become either excessively denominated as "Maya" (even when they do not self-identify as such), or excluded from the region's landscape altogether as they are deemed not sufficiently "Maya" (thus, not of interest that would benefit Mayanism). As researchers who have permanently lived in Chiapas (one with Mexican nationality, the other one with a permanent residence permit), it has often amazed us how even the Chiapanecan Indigenous people, who tend to refer to themselves as Tsotsiles or Tseltales, suddenly become "Maya" in English-language North American academic writing. The name "Maya" seems to be given when translated into English. What could this entail and implicate for the Maya Forest imaginaries?

One further aspect implicated in this tendency is the fact that even today, North American/European English-language publications often do not engage in deeper conversation with the scholarship produced in Mexico, Guatemala, and Belize, especially when written in Spanish. We believe that this kind of engagement would generate different kinds of academic writing and indicate improving epistemological justices. Many outsider authors have engaged in critical discussion concerning the coloniality of knowledge related to the Maya Forest. We ourselves, when writing these lines, are part of them; yet we also detect a tendency by which some authors have opted to tackle the challenge solely by criticizing other outsiders. While such critique is necessary and healthy, and again, we are implicated in this ourselves right here, it also occasionally paints the Maya Forest as a mere playground for outsiders. When reading this kind of literature, say, from Chiapas, it gives an impression of the Maya Forest as something only subject to foreigners or outsider influences and with no history, agency, or voice of its own. This happens, for example, when the Maya Forest is defined as an area composed of national parks merely copied from North American models, which ignores the countries' own rich conservation histories, or when the "conservationists" are outrightly designated as foreigners although many local inhabitants and community members are involved in important conservation endeavors and are thus conservationists themselves. A fine line exists between critically examining the considerable outsider influence in the Maya Forest and making it all about the outside. In this book, we also engage with this difficult balance.

Researchers in Mexico, Guatemala, and Belize have encountered different challenges, many of which share similar traits with those indicated above, but for different reasons. For example, the strong, rich, and critical field of Mexican Anthropology, which is predominant in the southern Mexican Maya Forest with its many research institutions (where we have been working), has traditionally included the political task of integrating Indigenous people as part of the nation (Bertely Busquets, 2015). The glorious Mayan past is a national pride and a component of nationalism, supported by research. At the same time, some Mexican anthropologists, who are proud to be in the field, support the Indigenous communities and struggles while considering themselves in service of them rather than of academia or the nation-state.

While Mexican academia is bigger and better funded, Guatemalan and Belizean academic research is more dependent on outsider funding and agendas. This also means that a great deal of research in Guatemala and Belize is conducted by outsiders, and thus, knowledges are taken outside. The Maya Forest countries involve important academic power structures. When available at the national scale, research funding tends to be restricted to local contexts and "national problems" without considering transboundary issues and processes. This means that those able to conduct broader territorial research also tend to be outsiders, with the exception of some binational or trinational academic collaborations that we are happy to have seen emerging.

In this context, the Maya Forest waterlands literature simultaneously reflects two tendencies: one that tends to magnify the challenges related to global power relations of knowledge production, and another that has enabled extraordinary efforts in bringing together scholars beyond national borders in a shared attempt to produce richer knowledge on the region. The two tendencies exist simultaneously and overlap: we all somehow relate to these two sides of the same coin.

Having said that, we first mention the book of Primack et al. (1998) as one of the most important ones concerning the Maya Forest. In our evaluation, it is one of the first compilations that defined and promoted the Maya Forest based on its original configuration and institutional collaboration. Yet, even more importantly, the book made a considerable effort in bringing together a regional approach to the Maya Forest, acknowledging the challenge for all scientists of limiting oneself to sites within the boundaries of one country because work transcending national borders is arduous for many reasons, and thus, few have attempted it. While most authors of the book contributed to writing from their respective corners of the Maya Forest, often bound to national borders, the book presents a broad range of histories and realities not limited to the "Maya". It focuses on the many conservation experiences, protected areas, wildlife, and also forestry and livelihoods.

In 2003, Gómez-Pompa et al. (2003), contributed by publishing another compilation focused on the human–wildland interfaces in the Maya lowland area. While not precisely mentioning the "Maya Forest", it centered on the Maya/forest interface and clearly represented the Mayanist Studies' development toward environmental and ecological history. The book dealt with different connections between biodiversity and agriculture, including plants and wildlife. In this compilation, water aspects were included, from hurricanes to seasonal forests, wetlands, and hydrogeology. In Mayanist Studies, the Maya region is often divided into three areas consisting of the Maya lowlands, highlands, and the Pacific. The lowlands, as indicated in Gómez-Pompa et al. (2003), are karst plains stretching from Campeche to Honduras and encompassing the whole Yucatán. The Maya highlands follow the mountain ranges from Tabasco to Honduras, whereas the Maya Pacific corresponds to the coastal plain running from Chiapas to El Salvador. As has been mentioned in this introduction, the concept of the Maya Forest brings together the Maya low- and highlands (sometimes also the Pacific), but limited to the borderlands of Guatemala, Belize, and the southern states of Mexico – in other words, contemporary trinational borderlands.

These compilations were followed by various publications that all used the concept of the Maya Forest for different purposes. Frequently cited is that of Nations (2006), which sought to provide a unified Maya Forest narrative, conservation experiences across the territory, and an introduction to the region addressed to potentially enthusiastic (eco)tourists. Although focused on building the regional Maya Forest perspective, the book was mainly structured on the "country-by-country" basis.

In 2015, Ford and Nigh (2015) published their extensive work on the *Maya Forest Gardens*, which sheds light on sustainable agroforestry through time and space. Agroforestry, indeed, has become a nascent research strand in the Maya Forest. This title combines long-term archaeological and anthropological research focused on environmental history. This was, again, followed by more critical works by Martínez (2016), who focused on conservation and the Mayas in Quintana Roo, and Ybarra (2018), who examined the Mayan land rights and conservation in northern Guatemala. Later, Laako et al. (2022) critically scrutinized the concept of the Maya Forest, tracing its origins as part of transboundary conservation. Many others have published on the Maya Forest. Most authors have used the term to refer to their subject area and hence have not exactly elaborated or defined it.

The rich scholarship and knowledge of the Maya Forest waterlands with its borders is, of course, not limited to these references. In the forthcoming chapters, we cite and discuss plenty more. When considering the nexus of our notions, we observe some important gaps and silences. A major one, which is also our focus in this book, relates to forests and waters. The discussion above shows more clearly how the Maya and the forest have been linked and researched. More often than not, these studies have neglected waters, although some research linking waters and the Maya does exist, especially in Archaeology (e.g., Chávez-Gurmán 2016; Walker 2016). The water-related scholarship has, nevertheless, also addressed the borders – for example, in terms of transboundary river basins, which also tend to include woodlands (e.g., Kauffer, 2011; 2018).

In terms of borders and borderlands, the scholarship has been particularly active in Mexico, given its many institutions established to explore the "forgotten Mexican southern borderlands" (e.g., Laako, 2016). These same narratives about the abandoned southern Mexico have now surfaced in the justification for the Maya Train. Thus, we perceive a contrast between the traditional, predominant borderlands' scholarship – especially tied to the southern Mexican "peripheries" – and the contemporary context of territorial, social, and environmental transformations. In this book, we challenge the above while, at the same time, humbly seeking to remediate the fact that most border studies in the Maya Forest center on the Mexican side. In this context, we would like to point out the works of Castillo et al. (2006) and Toussaint and Garzón (2023), who have conducted an analysis of the borderings between Mexico, Guatemala, and Belize, examining them from their different sides.

Despite being predominantly bound to national borders, the scholarship related to borders and borderlands connects to broader regional and territorial processes that importantly involve land issues. This is where the context-bound land ownership, land rights, and Indigenous/Mayan issues connect to broader geopolitical and territorial questions, making it difficult to locate the references in precise categories of "Maya", "border", and "land". Thus, in this strand, we mention various important references as examples involving those elements.

For instance, Martínez et al. (2023) have recently provided a compilation of social and natural scientific works that address the territorial and environmental implications of the Maya Train (see Map 0.2). While formally focused on the Mexican side, given the location of the train, many chapters included in the book contain a broader geopolitical analysis that integrates the broader Maya Forest (e.g., Ceceña & Prieto, 2023). Another example of this kind of geopolitical analysis related to the Yucatán Peninsula and mobilities is found in Prieto et al. (2021).

The above issues are, again, intimately linked to two influential aspects of the Maya Forest geopolitics that we mentioned at the beginning of this introduction: tourism and drug-cartel activity. Whereas the former has been subject to active scholarship (e.g., Fraga et al., 2015), few scholars have dared to take drug cartels as a research object. Yet, this topic is increasingly present in the field, endangering people, conservationists, and scholars alike. Unfortunately, it is as deeply entangled with the Maya Forest landscapes today as is tourism, and the two are intertwined. Concerning this topic, we in particular refer to the many publications authored by Devine (e.g., Devine et al., 2021), who explores drug-cartel activity in the Maya Biosphere Reserve and along the Guatemalan–Mexican borderlands. While our research has neither centered on tourism nor on drug-cartel activity, the two are constantly present in our fieldwork and writing.

The above-mentioned territorial changes, such as those implicated by the Maya Train, are intimately intertwined with the Maya and with the forest, yet not reflected upon in the predominant Maya Forest narratives. An extensive number of studies have dealt with different regional developments, lands, borders, and people, albeit bound to national borders. Yet, they have also spoken about broader regional and territorial developments and entanglements. Thus, we mention, for example, Wainwright (2008, 2009, 2015) and Shoman (2020), who have written about the Belizean Mayas, land right issues, and border politics. Schwartz (1990) and Grandia (2012) have addressed broader regional developments in Petén, Guatemala, including mobilities and livelihoods. In the case of the Guatemalan Maya Forest, the Maya Biosphere Reserve has attracted the interest of scholarship for a long time (e.g., Sundberg 2003). The same applies to Mexico, where, for example, the protected areas of the Lacandon rainforest have long stirred academic debates (e.g., Durand et al., 2014).

The historian de Vos (2002) has commented that in the scholarship on the Lacandon rainforest, geopolitical necessities have seemed to override any physical evidence. He made comments after observing how the Mexican researchers' writing on the Lacandon seemed to agree that the rainforest abruptly ends at the Usumacinta River, which is Mexico's contemporary border with Guatemala, although nature does not alter much across a river. This is precisely what Kauffer (2018) has called "methodological nationalism", which locks us into national borders as a point of departure. While we perfectly understand why and how this happens, as indicated in this section, we consider that the Maya Forest is also an invitation to think differently, and this is what *The Maya Forest Waterlands* sets out to do.

Next, we outline how we did this methodologically.

The Space and Place of Our Research and Writing

This book is a result of arduous work that has a space and a place. It has involved a space of writing and thinking since autumn 2022, when we first started to work on the book proposal. Since then, this book has involved bringing together two long-term research trajectories, many materials gathered and examined for this particular research, and an ongoing shared analysis that exceeded our existing categories and expectations. It has also involved a place – the Maya Forest Waterlands, with its many borders, locations, contexts, backgrounds, people, wildlife, natural elements, politics, and complex processes. It would have been easy to get lost, and we did so several times. However, what we found as a result of this triple somersault is presented in these pages.

The two research trajectories refer to the long-term work that we have been conducting on the Maya Forest waterlands for decades. While Kauffer has been particularly focused since 1993 on issues related to refugees and borders, as well as environmental politics and policies, particularly with regard to multiple water issues since 2003, Laako has been exploring various topics since 2005, from Indigenous movements and different activisms to decolonization and conservation politics. What we both share, in addition to an interest in environmental issues, is the study of geopolitics, political geographies, international relations, and borders. We are both political scientists by training, yet the context of southern Mexico, where we have been working and living for decades, has taught us about Anthropology.

Although we are most familiar with the Chiapanecan context, our research has always geared toward multiscale approaches and transboundarities. Laako expanded toward Tabasco and Guatemala in 2017, when she began her research on conservation politics. Since 2019, she has extended her work toward the other Mexican states of the Maya Forest and particularly Belize. Kauffer has periodically but constantly been involved in research related to the other Mexican states of the Maya Forest, as well as Guatemala and Belize. For example, she already conducted research in Guatemala and in the state of Campeche in the early 1990s, and at the end of that decade she led two international projects on transboundary river basins covering all of Central America, as well as a series of research projects including all Mexican bordering states.

In 2017, our shared work began in an interinstitutional and interdisciplinary megaproject on the transboundary Usumacinta River basin, with Kauffer, as the leader of the social and policy team, deepening her previous research on waters, and Laako on conservation. In 2019, Laako began her research on the Maya Forest, which then inspired her to expand toward the nexus of water/forest/conservation/borders and which, in many ways, culminated in this book. In our previous co-authored publications (Laako & Kauffer 2021 and 2022) we focused on policies and institutional developments, particularly in Mexico and Guatemala. We do not extensively repeat the contents of those publications in this manuscript and therefore recommend the reader to consult them for more precise details on policy developments and environmental institution-building related to the region.

Our fieldwork covers most national border areas of the Maya Forest Waterlands. This sounds like an easy statement but is far from it, implicating transboundary research in difficult locations and including administrative arrangements, long distances, and learning about processes in different countries, languages, cultures, contexts, backgrounds, and politics. Indeed, a major challenge when writing this book has been the balancing act between wishing to

Image 0.1 Hanna Laako and Edith Kauffer at the Rainy Mexican–Guatemalan border between Tabasco and Petén in July 2018
Source: Hanna Laako, 2018

give each issue and location the nuanced details they deserve and needing to capture broader tendencies in a vast territory overloaded with historical and political baggage; however, we would not want it to be any different. The balancing act between broader tendencies and more anthropological cases has allowed us to detect trajectories we would have missed otherwise. The transboundary work has confirmed the necessity to engage in fieldwork and triangulate sources as, time after time, we have seen that things are often not what they look like from afar. Moreover, field experiences may be misleading as they always offer a partial glimpse and thus need to be confirmed with other sources. Despite this partial glimpse, the attempt to engage in the field sheds light on aspects that could not be detected by distant mapmaking. This means that even what first may have seemed like "failed" fieldwork – meaning, for example, not being able to access a location or being faced with suspicion – ended up providing us with key information that we would not have discovered solely "online".

Considering the borders, we have many times, in different places, crossed the Mexican–Guatemalan border, particularly from Chiapas and Tabasco. We both have crossed borders between Mexico and Belize and between Belize and Guatemala. During the past years, Laako has, in particular, conducted fieldwork in the borderlands on the Belizean side, running from Chiquibul to the northern trinational corner with Guatemala and Mexico along the Río Hondo River. We both have conducted research in Petén, and Kauffer, in particular, in the

Northern Transversal Strip located south of the land border and across the departments of Huehuetenango, El Quiché, and Alta Verapaz. We have not been able to visit southeastern Petén around the Chiquibul-Maya Mountain borderlands. In 2023, Laako visited southern Belize; however, she was not able to access the border area along the Sarstoon-Temash National Park. Yet, our field observations on the different sides of the borders gave us a fuller panorama of the regional and national dynamics.

Our fieldwork and observation also include the most important protected areas of the Maya Forest Waterlands, running from the Mexican Montes Azules to Pantanos de Centla, from Laguna del Tigre to Calakmul, from Sian Ka'an to the Guatemalan Maya Forest in Petén, and in Belize, from Chiquibul to Río Bravo. The protected areas in the state of Yucatán and within the Great Barrier Reef in Belize are less familiar to us. We have also visited several communal, municipal/state, and private protected areas. Most fieldwork information used in this book derives from the years between 2017 and 2024. Our most detailed work related to communities derives from fieldwork along the Chiapas–Tabasco borderlands with Guatemala, which involved navigating for hours on their many rivers and regularly crossing the border at different points.

Our fieldwork has involved observation, interviews, discussions, and many kinds of other first-hand materials – among others, leaflets, recordings, notes, photos, reports, policies and maps. Moreover, the interviewees varied in their profile, although they were mostly conservationists, representatives of relevant governmental institutions and NGOs, and, above all, local inhabitants – men and women from communities, especially along the Mexican–Guatemalan borderlands and riverine contexts.

In addition, we have supported our findings with a broad range of literature reviews and other second-hand materials, the production and analysis of maps (as in this introduction), and also some newspaper articles and historical writings, such as those by archaeologists.

As indicated above, all this has made the writing challenging as, in addition to the broad fieldwork in a complex territory, the main objective of this book is to zoom in and out of several, pre-established, and highly abstract concepts, such as water, forest, Maya, border, and land. In the book, we subject them to critical scrutiny both theoretically and empirically. Thus, to be theoretically rigorous and empirically fair to both detailed nuances and broader tendencies, we needed to develop different ways of writing. For this reason, in the chapters that follow, the reader will find distinct solutions that combine empirical case studies, insights into broader observation and theoretical analysis.

The chapters may be read individually. For this reason, they may repeat some information provided elsewhere in this book. The chapters also include illustrative photos taken by us, which do not include many pictures with people. This is not because we have intentionally eliminated them from the landscape to convey empty wildernesses, but because we had to make these choices owing to complicated permit issues and ethical aspects. In what follows, the book is structured around the following chapters:

Chapter 1, *From Borderlands to Forest Waterlands,* is our theoretical framework, which dives deep into the concepts of borderlands, forest, water, and land. This chapter also outlines our main theoretical contributions by elucidating the corresponding theoretical discussions and perspectives, from borderlands to eco-borderlands, from ecoregions to biodiversity hotspots, and from the waterland nexus to rainforests. Despite being theoretical, it constantly refers to the Maya Forest, its origins, and its current definitions as a multiscale conservation space, based on active mappings.

Chapter 2, *Borderlands, the Maya, and the Creation of the Maya Forest*, continues from Chapter 1 by specifically addressing the concept of the Maya. Building on Borderlands Studies, and challenging the idea of the Maya Forest as a periphery, the chapter critically examines the three pillars of (re)production of the Maya: science, tourism, and the struggles of those who self-identify as Maya. The chapter particularly focuses on the coming-together of Archaeology and Ecology in building the Maya Forest. The chapter ends with two insights that illustrate the nexus of the Maya in its three pillars (borders, forests, and waters): one elucidating the history and contemporary context of the Usumacinta River and the other focused on the Columbia River Forest Reserve in the Belizean borderlands.

Chapter 3, *Rethinking Transboundarities: Connectivities of Water and Conservation in the Maya Forest Waterlands*, addresses another key angle to the Maya Forest waterlands: that of transboundarity. While often neglected from social scientific analysis, it has been actively promoted by conservationists as the main aim of the Maya Forest. In this chapter, we identify two different ways of understanding transboundarities: first, as a human-centered connectivity across international borders (as often defined by social scientists), and second, as a biodiversity-centered connectivity across political borders and ecosystems (as often defined by natural scientists). The two form their own kinds of borderlands. To explore these contradictions and challenges, we concentrate on water as a neglected and hidden issue that nonetheless allows us to put our finger on the multilayered connectivities of the Maya Forest waterlands. Thus, in this chapter, we draw from both types of transboundarities, first by addressing international rivers as political borders and then by elucidating the ecosystemic forest/water/land transboundarities and borderlands in conservation.

Chapter 4, *The Maya Forest Waterlands as Waterless Transboundary River Basins for Their Inhabitants*, discusses the tremendous contradictions between the extensive water resources elucidated in the case of two transboundary subbasins in the Maya Forest waterlands and the lack of water suffered by their inhabitants. In addition to these shared discrepancies, the populations that share the basins, but are located on different sides of the international borders show deep inequalities. In this way, the shared transboundary river basins, marked by borders, also highlight contrasting realities in the services accessible to their people. This shows that while the Maya Forest waterlands are rich in natural resources and biodiversity, their people often experience scarcity and contrasting inequalities – the many disconnections in contrast to the earlier connectivities.

Chapter 5, *Political Trails in the Maya Forest: Go-betweens, Curating, and Places-in-knots in Three Biological Stations*, dives deep into socio-political history and contemporary processes as embedded in three biological stations of the Maya Forest waterlands. Examining their origins, locations, and current challenges, the chapter presents a shared regional history of *chiclería*, that is, chewing-gum extraction. The chapter shows how *chiclería* is a neglected but formative transboundary history involving the forest and the Mayas, but also people and communities left out of the predominant Maya Forest landscapes. By studying the three stations and their embedded histories of *chiclería*, we find many shared geopolitical challenges, power relations, and symbiosis. The chapter shows how the Maya Forest unveils previously hidden shared histories and geopolitical dynamics involving the Mayas and forests.

We conclude by drawing together our main findings to show how the Maya Forest waterlands represent a call to rethink the transboundarities, distinct shared heritages, questionable perceptions of remoteness, complex geopolitical challenges, shapeshifting discrepancies, and always impactful scientific practice. By means of these chapters, we warmly invite the reader to share with us this exploratory path that reveals and explains the distinct political entanglements, shared spaces, and fluid borderings of the contemporary Maya Forest waterlands.

References

Ankersen, T. & Arriola, L. (2001). *At the Crossroads of Conservation and Development: The Challenge of the Maya Forest. A Report to the Coalición de la Selva Maya and the Tropical Ecosystem.* Directorate of the United States Man and Biosphere Program, University of Florida.

Bertely Busquets, M. (2015). De la antropología convencional a una praxis comprometida. Colaboración entre indígenas y no indígenas en un proyecto educativo para construir un mundo alterno desde Chiapas, México. In Solano, X.L., Leyva, X., Alonso, J., Hernández, R. A., Escobar, A., Köhler, A., Cumes, A., Sandoval, R., Speed, S., Blaser, M., Krotz, E., Piñacué, S., Nahuelpan, H., Macleod, M., Intzín, J.L., García, J.L., Báez, M., Bolaños, G., Restrepo, E., … de Sousa Santos, B. et al. (Eds). *Prácticas otras de conocimiento(s): Entre crisis, entre guerras* (Tomo I) (pp. 225–253). CLACSO.

Carabias, J., De la Maza, J. & Cadena, R. (Eds, 2015). *Conservación y desarrollo sustentable en la Selva Lacandona*. Natura Mexicana.

Castillo, M., Toussaint, M. & Vázquez, M. (2006). *Espacios Diversos, Historia en Común: México, Guatemala y Belice, la formación de una frontera*. Secretaría de Relaciones Exteriores de México.

Ceceña, A. & Prieto, S. (2023). Mirando al Sur: Megaproyectos, fronteras e (in)movilidades. In Martínez, E., Gasparello, G. & Díaz, M. (Eds). *Territorios mayas en el paso del tren: Riesgos previsibles y posturas independientes sobre el Tren Maya* (pp. 25–51). Bajo Tierra Ediciones.

Chávez-Guzmán, M. (Ed, 2016). *El manejo del agua a través del tiempo en la península de Yucatán*. Universidad Autónoma de Yucatán, Centro de Investigaciones Regionales Dr. Hideyo Noguchi, Fundación Gonzalo Río Arronte IAP and Consejo de Cuenca de la Península de Yucatán.

CONAP-DGPCN. (2019). *Plan Maestro de los Complejos III y IV de las Áreas Protegidas del Sureste de Petén*. Tercera Edición - Actualización. Asociación Balam y Fondo Nacional para la Conservación de la Naturaleza - FONACON. Guatemala.

Devine, J., Currit N., Reygadas Y., Liller, L., & Allen, G. (2021). Drug trafficking, cattle ranching and land use change in Guatemala's Maya Biosphere. *Land Use Policy*, 95, 104578.

de Vos, J. (2002). *Una tierra para sembrar sueños: Historia reciente de la Selva Lacandona, 1950–2000*. CIESAS.

Devine, J. (2016). Contesting global heritage in the Chicle Workers' Museum. *Latin American Research Review*, 51(3), 101–122.

Durand, L., Figueroa, F., & Trench, T. (2014). Inclusion and exclusion in the participation strategies in the Montes Azules Biosphere Reserve, Chiapas, Mexico. *Conservation and Society*, 12(2), 175–189.

Fisher, C. & Chase, A. (Eds, 2021). *Maya Anthropological Archaeology*. MDPI Books.

Ford, A., & Nigh R. (2015). *The Maya Forest Garden: Eight Millennia of Sustainable Cultivation of the Tropical Woodlands*. Routledge.

Fraga, J., Khafash, L. & Córdoba, J. (Eds, 2015). *Turismo y ocio: Reflexiones sobre el Caribe Mexicano*. PASOS: Revista de Turismo y Patromonio Cultural.

García, G. & Secaira, F. (Eds, 2006). *Una visión para el futuro: Cartografía de las Selvas Mayas, Zoque y Olmeca*. Infoterra Editores.

Gómez-Pompa, A., Allen, M., Fedick, S. & Jímenez-Osornio, J. (Eds, 2003). *The lowland Maya area: Three millennia at the human-wildland interface*. CRC Press.

Grandia, L. (2012). *Enclosed: conservation, cattle and commerce among the Q'eqchi' Maya Lowlanders*. University of Washington Press.

Global Conservation. (2024). Greater Belize Maya Forest. https://globalconservation.org/projects/greater-belize-maya-forest/ (Consulted in March 2024).

Harris, L.M., & Hazen, H.D. (2015). Power of maps: (Counter) mapping for conservation. *ACME: An International Journal for Critical Geographies*, 4(1), 99–130.

Hostettler, U. (2004). Ethnographic perspectives in Maya studies: Trends in writing about Mayas in Guatemala, Mexico and Belize. *Ethnohistory*, 51(2), 435–444.

Ingold, T. (2000). *The perception of the environment: Essays on livelihood, dwelling and skill*. Routledge.

Kauffer, E. (Ed, 2011). *Entre manantiales y ríos desatados: paradojas de las hidropolíticas fronterizas (México-Guatemala)*. CIESAS.

Kauffer, E. (Ed, 2018). *Cuencas transfronterizas: La apertura de la presa del nacionalismo metodológico*. CIESAS.

Laako, H., & Kauffer, E. (2022). Between colonizing waters and extracting forest fronts: Entangled eco-frontiers in the Usumacinta River basin. *Political Geography*, 96, 102566.

Laako, H., & Kauffer, E. (2021). Conservation in the frontier: Negotiating ownerships of nature at the Mexican southern border. *Journal of Latin American Geography*, 20 (3), 40–69.

Laako, H. (2016). Decolonizing vision on borderlands: The Mexican southern borderlands in critical review. *Globalizations*, 13 (2), 173–187.

Laako, H., Pliego, E., Ramos, D. and Márquez, B. (2022). Transboundary conservation and nature states in the Maya Forest: International relations, challenged. *Globalizations*, 19(8), 1288–1310.

Leyva, X., & Burguete, A. (2007). *La remunicipalización de Chiapas: Lo político y la política en tiempos de contrainsurgencia*. CIESAS.

Loucky, J., & Moors, M. (Eds, 2000). *The Maya Diaspora: Guatemalan Roots, New American Lives*. Temple University Press.

Lucchesi, A.H. (2018). "Indians don't make maps": Indigenous cartographic traditions and innovations. *American Indian Culture and Research Journal*, 42(3).

Martínez, E., Gasparello, G., & Díaz, M. (Eds, 2023). *Territorios mayas en el paso del tren: Riesgos previsibles y posturas independientes sobre el Tren Maya*. Bajo Tierra Ediciones.

Martínez-Reyes, J. (2016). *Moral ecology of a forest: The nature industry and Maya post-conservation*. University of Arizona Press.

McGurk, T.J., & Caquard, S. (2020). To what extent can online mapping be decolonial? A journey throughout Indigenous cartography in Canada. *The Canadian Geographer / Le Géographe Canadien*, 64, 49–64.

McSweeney, K., Richani, N., Pearson, Z., Devine, J., & Wrathall, D.J. (2017). Why do narcos invest in rural land? *Journal of Latin American Geography*, 3–29.

Mejía-Ortiz, L.M., Sprouse, P., Tejeda-Mazariegos, J.C., Valladarez, J., Frausto-Martínez, O., Collantes-Chávez-Costa, A.L., Ruíz-Cancino, G., & Yáñez, G. (2021). Tropical subterranean ecosystems in Mexico, Guatemala and Belize: a review of aquatic biodiversity and their ecological aspects. In Hufnagel, L. (Ed), *Natural History and Ecology of Mexico and Central America*. IntechOpen. https://doi.org/10.5772/. 97694

Metcalf, A. (2020). *Mapping an Atlantic world, circa 1500*. John Hopkins University Press.

Miller, S.W. (2007). *An environmental history of Latin America*. Cambridge University Press.

Nations, J. (2006). *The Maya tropical forest: People, parks, and ancient cities*. University of Texas Press, Austin.

Portillo-Quintero, C., Hernandez-Stefanoni, J.L., & Dupuy, J.M. (2023). Forest clearing dynamics and its relation to remotely sensed carbon density and plant species diversity in the Puuc Biocultural State Reserve, Mexico. *Remote Sensing*, 15, 3445.

Pozo, C., Armijo, N., & Calmé, S. (Eds, 2011). *Riqueza biológica de Quintana Roo: Un análisis para su conservación, III tomos*. CONABIO, ECOSUR, Gobierno de Estado de Quintana Roo, y Programa de Pequeñas Donaciones.

Prado, F. (2012). The fringes of empire: Recent scholarship on colonial frontiers and borderlands in Latin America. *History Compass*, 10(4), 318–333.

Prieto, S., & Benítez, J. (2021). Vías de comunicación, reordenamiento territorial y reconfiguración de las (in)movilidades humanas en la península de Yucatán y la frontera sur de México. In Benítez, J., & Escalona-Segura, G. (Eds). *Impacto de las vías de comunicación sobre la fauna silvestre en áreas protegidas. Estudios de caso para el sureste de México* (pp. 166–204). ECOSUR.

Primack, R., Bray D., Galletti H., & Ponciano I. (Eds, 1998). *Timber, tourists, and temples: Conservation and development in the Maya Forest of Belize, Guatemala and Mexico*. Island Press.

Rahder, M. (2020). *An ecology of knowledges: Fear, love and technoscience in Guatemalan forest conservation*. Duke University Press.

Ramsar Sites Information Service (2009). *Anillo de Cenotes*. https://rsis.ramsar.org/ris/2043 (Consulted in March 2024).

Sarstoon-Temash National Park: Transcript of Stakeholders' Workshop (1998), held on February 22, 1997 at the Community Center, Barranco Village, Toledo District, Belize. Producciones de la Hamaca Caye Caulker & Community Conservation Consultants Publishing. Belize Archives and Records Service.

Schwartz, N. (1990). *Forest society: A social history of Petén, Guatemala*. University of Pennsylvania Press.

SEGOB, DOF 30.11.2018. Acuerdo por el que se da a conocer el Resumen del Programa de Manejo del Área Natural Protegida con categoría de Reserva de la Biosfera la región conocida como Caribe Mexicano. https://dof.gob.mx/nota_detalle.php?codigo=5545531&fecha=30/11/2018#gsc.tab=0 (Consulted in March 2024).

Shoman, A. (2020). La resolución de conflictos territoriales: El caso del Proceso de Conciliación Belice-Guatemala (2000–2020). *Anuario de Estudios Centroamericanos*, 46, https://doi.org/10.15517/aeca.v46i0.45094.

Sletto, B., Bryan, J., Wagner, A., & Hale, C. (Eds, 2020). *Radical cartographies: Participatory mapmaking from Latin America*. University of Texas Press.

Sundberg J. (2003). Conservation and democratization: Constituting citizenship in the Maya Biosphere Reserve, Guatemala. *Political Geography*, 22(7), 715–740.

Toussaint, M. (2009). Entre los vecinos y los imperios: El papel de Belice en la geopolítica regional. *Tzintzun: Revista de Estudios Históricos*, 50, 105–129.

Toussaint, M., & Garzón, M. (Eds, 2023). *Dynamics and Conflicts in a Cross-Border Region: Mexico, Guatemala and Belize*. Routledge.

UNESCO Man and Biosphere Programme and lists of biosphere reserves:https://www.unesco.org/en/mab (consulted in March 2024).

Viquiera, J., & Ruz, M. (Eds, 2004). *Chiapas: Los rumbos de otra historia*. UNAM and CIESAS.

Wainwright, J. (2008). *Decolonizing development: Colonial power and the Maya*. Blackwell Publishing.

Wainwright, J., & Bryan, J. (2009). Cartography, territory, property: postcolonial reflections on indigenous counter-mapping in Nicaragua and Belize. *Cultural Geographies*, 16(2), 153–178.

Wainwright, J., Jiang, S., Mercer, K., & Liu, D. (2015). The political ecology of a highway through Belize's forested borderlands. *Environment and planning A: Economy and space*, 47(4), 833–849.

Walker, D. (Ed, 2016). *Perspectives on the ancient Maya in Chetumal Bay*. University Press of Florida.

World Bank (2022). Terrestrial protected areas. World Database of Protected Areas. https://data.worldbank.org/indicator/ER.LND.PTLD.ZS?end=2022&skipRedirection=true&start=2016&view=map (consulted in July 2024).

Ybarra, M. (2018). *Green wars: Conservation and decolonization in the Maya Forest*. University of California Press.

1 From Borderlands to Forest Waterlands

Hanna Laako and Edith Kauffer

Introduction

In this chapter, we shed light on our forest waterlands perspective to be employed and explored in the book. Forest waterlands fundamentally build on Borderlands Studies. However, we simultaneously examine bridges between the multidisciplinary Borderlands Studies and environmental social sciences, which connect to the fields of Tropical Ecology and Conservation Biology that seek the integration between the cultural/political and ecological. We argue that these paradigms share more than what appears at first glance. They should engage more with each other as they are all interested in the same, previously remote, peripheral fringes of the world: the borderlands that are now recognized as biodiversity hotspots for conservation are also places of encounter and intermingling with *longue durée* history and topical contemporary dynamics, particularly related to Indigenous territories and livelihoods. Sometimes, as in our case of the Maya Forest, they are also transboundary regions between countries. In this chapter our underlying inquiry is the following: what happens to borderlands when they are shaped by contemporary conservation and environmental management? How do these processes (and corresponding disciplines) mold Borderlands Studies and its conceptual work and vice versa?

This chapter examines and bridges these academic discussions while addressing the increasing need to analyze the cultural/political and ecological borderlands, or eco-borderlands, considering recent developments and dynamics in global conservation and natural resource management, which have transformed and rekindled the interest in distinct borderlands. These may be borderlands that, so far, have been mainly shaped – and analyzed – as Indigenous territories encountering empires or colonialism, or extractive appropriation and resource management. Yet, there are different kinds of borderlands: forgotten peripheral borderlands left behind by development and modernization; borderlands previously neglected by Borderlands Studies scholars themselves, as they were perceived as mere bushlands with no human population (i.e., the peripheral fringe of Borderlands Studies); borderlands that are rural areas whose inhabitants hope for alternative ways of life; borderlands now subject to environmental rescue efforts, in the global spotlight of ecosystem concern and consciousness,

DOI: 10.4324/9781003429050-2

which simultaneously question the remoteness of these backlands; and today's borderlands, bordered by protected lands or international borders, which may be questioned by ecological actors seeking to unify ecoregions, thus forming new kinds of bordered lands. How should we conceptualize borderlands in the light of these developments? What can Borderlands Studies contribute to the contemporary understanding of these edges of empires and places of encounters, not just of humans, but of species and landscapes?

These are the questions we examine in this chapter while building and explaining our analytical perspective of "forest waterlands", which derives from borderlands but also encompasses the critical approach to the forest–water–land nexus as politically separated yet practically unified categories. Building on environmental social sciences, we define borderlands as fluid edges including the cultural, political, and ecological. We simultaneously address the construction of ecoregions by conservation biologists and the debates related to the intermittent and changing nature of categories such as water, land, and forest. This analytical perspective is not concerned with fixed, closed categories; rather, it allows to ponder on what happens in the world's borderlands, when they are shaped by such processes as conservation and environmental management.

To this end, in this chapter, we first shed light on the main developments in Borderlands Studies, explaining what they are about and their relationship to natural environments. We then proceed to discuss eco-borderlands as fluid edges; we explore how environmental social sciences can contribute to the understanding of borderlands and how Borderlands Studies has specifically addressed ecological aspects. Finally, we arrive at our main conceptual perspective of "forest waterlands". In this final section, we explain how Tropical Ecology and Conservation Biology have contributed to shaping contemporary borderlands, as well as the need to critically explore the ecological categories in terms of borderlands definitions.

Borderlands Studies: On the Things Left Underfoot

Borderlands Studies was born as a critical paradigm centered on counter-narratives and supposed geographical margins or remote areas. These are also the reasons why this book builds on Borderlands Studies, particularly seeking to contribute to this field with the introduction of the concept of forest waterlands, critically exploring the environmental dimensions of borderlands.

In the Americas, Borderlands Studies forms its own multidisciplinary field. It was born in the 1920s as a critique of the then-predominant frontier concept (e.g., Hämäläinen & Truett, 2011). Both concepts – frontier and borderlands – refer to peripheral areas, fringes, or edges of empires. However, the frontier concept was increasingly criticized as deeply colonial, expansionist, and centered on conquest. We have previously dealt with the frontier concept and its current re-emergence as eco-frontiers (e.g., Laako & Kauffer, 2021, 2022), as well as its entangled and antagonistic relationship with borderlands and nature (e.g., Laako, 2023). In particular, the borderlands literature has taken critical distance from the colonial, binary categories of the frontier and defined borderlands as entangled, open-ended histories set in peripheral and bordered lands, where different types of negotiations and open-ended encounters take place. Borderlands have been studied from their own perspective and in relation to the supposed power centers.

Borderlands Studies grew significantly as a field in the 1990s. The field also underwent a so-called cultural turn influenced by Indigenous/Native Studies. According to Radding (2014), Borderlands Studies, which has been particularly rich in fields related to Global History, Social Anthropology, Human Geography, Arts and Regional/Area Studies, has focused on the entanglements between settlers, states, and imperial powers with nature and Indigenous people. These have been issues of particular interest in the Americas. Radding (2014) argues that Europeans have arrived late to Borderlands Studies for many reasons: the conceptual frames have been different, and borders and state-centrism play a different role in European countries, where there have been fewer Indigenous encounters and large-scale settlement projects than in the Americas. Borderlands Studies has thus been (perhaps overly) conditioned by the North American example and, to some extent, by its counterpart: the frontier.

However, this might be changing with the new type of borderlands writing from the *Critical Norths* – the Arctic, the Sami people, and hybrid, peripheral North-Karelian borderlands (e.g., Kullaa et al., 2022; Lehtinen, 2006; Ray & Maier, 2017) – whereas, in the so-called Global South, the concept of the frontier, which is associated more clearly with resource extraction, settler advancement, and de/re-territorialization, is still gaining ground (e.g., Laako & Kauffer, 2021). In Latin America, environmental historians have been particularly active in employing the borderlands perspective (e.g., Prado, 2012).

As this description suggests, the borderlands approach is also different from Border Studies, which is predominantly focused on the construction, development, and dynamics of international legal borders. While Borderlands Studies also involves international borders and thus shares some key elements with Border Studies, the borderlands discussed here are not imposed from above as imperial or national borderlines (Readman et al., 2014). Rather, we build on the literature that approaches borderlands as contested territories vitally shaped by local experiences, and as places of encounter and entanglement. This is distinct from the usual frontier literature and is not reduced to areas surrounding international borders (e.g., Anzaldúa, 1987). Our focus is on the strand that takes critical distance from the frontier concept, which is considered different but related.

Our strand of Borderlands Studies is a critical paradigm made up of counternarratives and focused on the world's remote places. We do not intend to develop, in detail, the shared and contentious conceptual history between frontiers and borderlands that we have already addressed elsewhere (e.g., Laako & Kauffer, 2021, 2022). Our conceptual history of borderlands situates the field as a critical paradigm composed of counternarratives, seeking an alternative angle to the predominant mainstream power relations and histories. Whereas the frontier concept has tended to use the image of an advancing front, conquering wildernesses by means of pioneers, settler colonialism, and nation- and state-building, as a way of explaining particular power dynamics, the borderlands approach emphasizes those aspects that the frontier has left underfoot.

Herbert Eugene Bolton, often cited as the founding father of Borderlands Studies, was particularly interested in those lost worlds left behind by the advancing frontiers (e.g., Hämäläinen & Truett, 2011) - rather than mythological wilderness, he saw contested territories. In other words, Borderlands Studies comprises counternarratives that recognize both the power structures and the continuity of Indigenous territories. In this sense, the borderlands approach shares aspects with both Subaltern Studies and Social and Economic History of ordinary people. It takes critical distance from state-centrism and borders as mere dividing lines and centers around contact zones and transculturation, diaspora, contiguous spaces (both horizontal/geographical and vertical/socio-structural), interfaces, and interaction between different peoples, empires, and nations.

Borderlands Studies also suggests geographical and illustrative places: edges of empires, fringes, peripheries, and margins, which may be international borders but are not limited to them. Here, Borderlands Studies can be perceived as sharing theories with critical International Relations on global structures, world order, and systems analysis, as well as postcolonial perspectives.

Levin Rojo and Radding (2019) have argued that borderlands are lived-in spaces rather than boundaries dividing social, economic, or political entities, and that these spaces become borderlands when (1) two or more spheres of hegemony limit and overlap each other frequently, claiming rights to resources and the control of people; (2) two or more groups of people with different cultures and modes of life intermingle; and (3) prevailing ecological conditions represent a challenge to particular forms of human habitation, thus conditioning livelihoods, or where natural environments undergo modifications resulting from the productive and settlement practices of the people inhabiting them. This definition emphasizes dynamic transitions and intermingling in power relations, yet it does not indicate a closed history of exploitation and appropriation. Borderlands are concerned not only with problems and conflicts, but also with middle grounds of *go-betweens* (Metcalf, 2005; Prado, 2012).

We, however, find it necessary to emphasize another layer in this definition of borderlands: their supposed remoteness or marginality, as the concept of "borderlands" not only denotes extraction, settlement, conquest, and state-building, but also the territories overshadowed by these, i.e., left underfoot by these actions. In other words, their margins, which may be multispecies landscapes, but also capitalist ruins (Lowenhaupt Tsing, 2015) or places that appeared distant in the past but have now become accessible (Vannini, 2024). Therefore, remoteness is relational and situational rather than an absolute or objective time-distance condition: it is not centered on stable physical distances, but on an emergent and shifting configuration of multiple and constantly evolving, temporary connections, disconnections, entanglements, and mobilities resulting from different social, cultural, economic, and geopolitical forces (Vannini,

2024). Thus, borderlands implicate a perception based on a place or a location at a distance or at the edges of power. Sometimes, peripheries are built or perceived in unlikely places. The perception depends on who is looking and from where: from self-defined peripheries, where marginality is felt with the absence of connections or development, to mention an example, or from self-defined cores, which situate the perceived leftover regions in a peripheral position. Borderlands are often rural or transitory places left behind by development or subject to renewed interest by modernizing projects. The focus on the situatedness and relational understanding of concepts and intersectionality derives from Gender Studies.

Borderlands Studies also connects to what could be called the *Geography of the Margins*, proposed by Mustonen et al. (2017), which is born at and pushed to the edges. However, the margins are also places that allow the emergence of critical voices and alternative livelihoods not possible elsewhere, perhaps precisely because they have been overshadowed or unexplored; they are places where different life forms may still have space to exist as they are only vaguely attached to the mainstream. Thus, the peripheries, rural areas, and margins are not only objects of external influences, but actors in their own right and with their own shared continuities. As Mustonen et al. (2017) have suggested, the margins are mirrors of power while also forming their own cores; therefore, they are intimately connected to power relations and to the narratives and interpretations of the core. Nonetheless, as also noted by Hämäläinen and Truett (2011), it would be a mistake to assume that margins and borderlands are automatically what the cores or power centers are not. Margins, peripheries, and borderlands are not necessarily places of the extraordinary, but often of the regular and ordinary – and sometimes also of decay and parochiality.

The understanding of borderlands as margins, peripheries, or rural or transitory areas left underfoot is not straightforward; rather, it invites further curiosities. A tension is clearly at play between defining borderlands as a lived-in space, where two or more hegemonies or groups of people entangle, and a peripheral margin, left underfoot or a forgotten area in development, which forms a particular entanglement with a power center, core, or mainstream. Thus, borderlands can be perceived both as a central or a core place for encounter and exchange that creates its own dynamics, and as a fringe that builds its own dynamics owing to a changing relationship with a supposed core or central area, which can also include abandonment and exclusion. These critical questions about the rural, remote, and peripheral and their complex layers are the focus of our inquiry into borderlands, and at the center of our understanding of the concept. Rather than forming a fixed category of remoteness, ruralness, or marginality, we explore those meanings in the whole of this book.

These ideas coincide with the conceptual understanding of the Spanish word *frontera*: in addition to alluding to the border/frontier, it refers to edges, wilderness, disputed territories, and open areas for seeking resources and trade opportunities (Prado, 2012). *Frontera* also refers to unpopulated lands, uncultivated wilderness, colonial peripheries, deserted or unexplored lands, rural areas, and backlands – hence, both a place for encounter and a fringe to a core. In English, the terms "border", "boundary", and "frontier" are often used synonymously (Djordje-Stojanovic, 2018, p. 110). However, as shown in our discussion of borderlands, concepts – even paradigms – have different meanings. According to Djordje-Stojanociv (2018), a boundary bounds or limits something and is more comprehensive than the specific border and frontier; borders can be legal lines, but boundaries are more flexible and may also be sociocultural, economic, geopolitical, and biophysical. Frontiers, again, are clearly centrifugal.

Our fundamental focus, however, is on how natural environments relate to borderlands. As suggested above, Borderlands Studies is a field positioned at the crossroads of many critical theories and disciplines belonging to the social sciences and humanities; but, what about environmental social sciences? In which ways could Borderlands Studies speak to and communicate with environmental social sciences?

According to Levin Rojo and Radding (2019), as borderlands are socially produced spaces and places, the natural conditions of climate, topography, and hydrology cannot create borderlands in and of themselves, despite being integral components of them. They have argued that, while Borderlands Studies arises with a strong environmental component (such as the Arctic tundra, Great Plains, and tropical lowlands of Amazonia), borderlands cannot exist in the absence of substantial (physical/local?) human exchange and interaction. Readman et al. (2014) have also stated that "(…) This is because borderlands cannot exist in the absence of significant human exchange and interaction: as Antarctica is very largely uninhabited and has no Indigenous population, its borders do not define borderlands" (p. 3).

We partially challenge this understanding. We agree that cross-cultural interaction and (human) populations are key to understanding borderlands. However, based on the above discussion about borderlands as remote edges, we also maintain that an ecological region, an abandoned rural area, the so-called wilderness, a non-indigenous forest plantation, and even Antarctica, can form a borderland when people perceive them as such – from a distance, but indicating interaction in the creation of a particular fringe. A place such as Antarctica can become a borderland – an edge – by extending beyond other inhabited, built, and colonized areas, yet being subject to curiosity, and as backlands subject to research and exploration. Moreover, our notion of borderlands is based on the premise of Hämäläinen and Truett (2011), who argue that Borderlands Studies was born to take critical distance from the colonial binary categories, such as conquerors/conquered, imperialists/Indigenous peoples, insiders/outsiders, and state/community; to follow this line, we might add also nature/human.

In this book, we explore the ways in which borderlands are eco-borderlands that may have been previously marginalized, abandoned, or remote backlands, but which have now become lands full of biodiversity to be protected or lands with natural resources to be managed – with or without inhabiting and migrated/evicted human populations and livelihoods. We also examine all the tensions that these processes create, which simultaneously transform the territories, natural environments, and populations in question. We address how borderlands are entangled with natural environments and how the conceptualization of borderlands may be developed toward new ways of integrating natural environments into their analytical frame, beyond playing the role of a frozen extra in the background scenery. Thus, the concept we use in this title – *forest waterlands* – primarily derives from, builds on, and contributes to Borderlands Studies, and we do so both by exploring the links between Borderlands Studies and environmental social sciences and by means of our transboundary Maya Forest waterlands case. We also address other related conceptual discussions, such as ecoregions and wetlands, generally used by ecologists with bio-geographical connotations and originating from natural sciences, and the water–land nexus proposed by Cortesi and Camargo (2022), which seeks to question the conceptual divides between waters and lands.

In the following section, we first address the notion of eco-borderlands as entangled, fluid edges. Subsequently, we define the related terminology, such as ecoregions, and its relation to the conceptual frame proposed here. Finally, we conceptualize forest waterlands.

Eco-Borderlands as Entangled, Fluid Edges: "Making Worlds Is Not Limited to Humans"

Lowenhaupt Tsing (2015) has argued that "[t]he making worlds is not limited to humans" (p. 22), and the same could be said of borderlands. Frontiers and borderlands are concepts with a shared interest in the world's edges, fringes, and outermost regions, and they both involve natural environments: natural elements, ecosystems, geographies, and wildlife. However, the frontier literature has tended to view natural environments from the standpoint of conquest, extraction, and appropriation of resources, whereas the borderlands literature has viewed nature more as a living space and as part of cultural interaction and heritage. Nevertheless, in both literatures, natural environments often remain static landscapes rather than gaining momentum as supporting or main actors. Lowenhaupt Tsing (2015) has changed this by treating pine trees and woods as active historical actors in her study of the global *matsutake* chain, which is a pioneering piece of work in the social-sciences strand of Multispecies Studies (see also Haraway, 2007). While, in this book, we cannot consistently employ such a multispecies perspective (nor is our approach focused primarily on *species)*, our notions of eco-borderlands and forest waterlands are influenced by this literature, which seeks to create an equilibrium between human and natural worlds while taking critical distance from the human-dominated analysis in social sciences (e.g., Laako, 2023; Laako et al., 2022).

The latter is important for several reasons. First, the contemporary understanding of the Anthropocene, implicating changes and disappearances of species and environments, is crucial for such a critical crossroads paradigm as Borderlands Studies. Simultaneously, the Anthropocene incites a change of perspective between the nature/culture binary typical of such a field as Borderlands Studies.

Second, supposedly peripheral and often biodiversity-rich regions have become the subject of increasing ecological concerns and actions: renewed attention has been paid to remote borderlands, now viewed as forests and waters filled with biodiversity or the locations of so-called catastrophes or disasters, such as wildfires and floods. To differing degrees, these environmental concerns have brought about, among others, bordered protected areas, environmental policies, conservation organizations and projects, rangers, environmental collaboration with communities as well as their resistance and alliance building, environmentally inspired border-crossing activities, natural resource governance, and incentives of all kinds, which shape these borderlands. Some of these processes also suggest significant tourist operations based on the attraction of remote, rare landscapes and wildernesses. New labor markets and jobs have been created around green transitions and economies: many projects bring – albeit precarious – employment as wildlife guides, nature photographers and cartographers in tourism or in international nongovernmental organizations, but with the involvement of many local and Indigenous communities as well as settlers (e.g., Neimark et al., 2020). Borderlands also involve research and science activities, or community forestry. Consequently, the geographies and cultures of these borderlands change together with the changing nature–culture relationship. These should be Borderlands Studies *par excellence*, yet our finding is that the consideration of what *ecological borderlands* could entail between environmental social sciences and Borderlands Studies is still in its nascent stages, perhaps owing to the continued dominance of cultural and Indigenous approaches, and a strong focus on borderlands history. Notably, Indigenous perspectives and natural environments should not be considered mutually exclusive.

Political Ecology has actively addressed the impacts and colonial nature of contemporary conservation and environmental management in Indigenous territories and other livelihoods (e.g., Carpenter, 2020; Duffy, 2022; Dudley & Stolton, 2020; Ybarra, 2018). It has examined neglected aspects of contemporary environmental politics in remote places and in the borderlands of protected areas; however, these lands are often characterized by a frontier-type approach focused on conquest and appropriation, which differs from the Borderlands Studies' emphasis on understanding entanglements and go-betweens (e.g., Laako, 2023; Laako & Kauffer, 2021). In other words, these studies risk presenting people as hapless victims of advancing state-building by means of environmental policies, or of international interventions by the great conservation organizations. Perhaps for the same reason, political ecologists have been keen on developing and rethinking the concept of frontiers or borders, but not as much Borderlands Studies (e.g., Büscher, 2013; Peluso & Lund, 2013; Ramutsindela, 2014).

Borderlands have been used in environmental social sciences to refer to the space and relationship between human populations and wildlife (e.g., Emel &

Wolch, 1998; Johansson, 2008). In other words, borderlands are the space or place where humans and nature, species, and local people meet and interact (Salisbury & Fagan, 2013; Zimmerer, 2007). These types of ecological borderlands could then refer to those edges where urban environments recede and natural environments begin, or the precise margin where the two meet, despite not belonging clearly to either world. In this case, we would be exploring rural areas, secondary forests, plantations, or outskirts of urban centers, to mention some examples. Nevertheless, these studies deriving from environmental social sciences in general and using the concept of borderlands do not really engage with the Borderlands Studies literature discussed here. Borderlands are merely an operationalized term to refer to an encounter between human populations or built-up areas and wildlife species/natural environments. Given that human impact is everywhere, the critical Anthropocene literature can, of course, ask: *are there any eco-borderlands left, or are eco-borderlands everywhere?* Again, we return to the tension between defining borderlands as middle ground and as margin. In the Anthropocene, are eco-borderlands, understood as middle grounds, everywhere, or are eco-borderlands getting smaller?

According to Hämäläinen (2010), Borderlands Studies, in its critical history of colonialism, first went through a so-called biological turn by exploring epidemiology and ecological imperialism during the conquest and colonization. These bio-histories were also criticized for their biological determinism. Hämäläinen (2010) himself has developed a notion of *ecological safety valve* that refers to the Indigenous people' fight against ecological destruction and the loss of lands, territories, and natural resources. This could be an interesting point of departure for critical political ecologists examining contemporary Indigenous livelihoods and lands.

With regard to environmental social sciences, or humanities, and engagement with the strand of Borderlands Studies discussed here, environmental historians have been particularly active in the integration of natural environments with borderlands. Prado (2012) has argued that new Environmental History has encouraged research on how nature transforms societies and nation-states and how certain borderlands, as interstices, have also changed social and ecological landscapes. This form of environmentally and geographically focused Borderlands Studies takes critical distance, again, from mainstream narratives and views that have been overshadowed by the predominant US–Mexican borderlands, as well as from Anglo- and Central European environmental theories that have also for a long time defined research concerning other regions of the world. Lehtinen (2006), in his community-based research, has for example addressed the Fennoscandian Green Belt as a geopolitical, societal, and environmental hybrid, which differs from the way North America or continental Europe understand conservation histories. Similarly, Wakild (2013) has studied conservation history in Mexico, Chile, and Argentina, noting how nature conservation has been connected to issues of environmental justice and land rights, in contrast to North American idea(l)s of wildernesses and national parks. Simultaneously, she has questioned whether conservation in Latin

America can be reduced to an entirely elitist project, when in 1940s Mexico, the Lázaro Cárdenas presidency focused on land reform and also created 40 national parks as part of its social policy addressed at marginalized people. We (Laako & Kauffer, 2022) have also found that the relationships between conservationists and local communities (Indigenous and non-Indigenous) are so entangled and complex that they cannot be reduced to a simple insider–outsider binary. Andersson et al. (2021) have concluded that the binary category of considering conservation simply as a source of conflicts within Indigenous communities has been necessary, but has also harmed the understanding and actual collaboration between conservationists and Indigenous people, some of whom are actually both.

Prado's (2012) research is part of the strand of environmental and borderlands history focused on the Transatlantic Frame, which considers the Atlantic shores, sheds, and basins as its own oceanic borderland based on the shared cultural history of slavery. This literature asks: *can an ocean be a borderland?* (Guidotti-Hernández, 2016; Metcalf, 2005). Similarly, the Pacific Rim countries have been studied as their own ocean-born borderland of encounter and interaction (e.g., Crewe, 2017). Crewe (2017) has argued that the study of the global history of connectivity, such as that of the Pacific Rim, allows critical distance from both inward-looking Area Studies and Eurocentric narratives, which have often positioned Latin America as peripheral. Other examples include the previously mentioned Arctic North, the transboundary Karelian borderland between Finland and Russia, the Baltic Sea borderlands, the Sápmi (transboundary Sámi homeland), and the Amazon. The Amazon represents an interesting case study owing to its global role as "the lungs of the earth" (Garfield, 2012). It is an example of a borderland defined by a rainforest, whose inhabitants engage in their own interactions with nature and each other, constituting a particular space; it is also formed of a remote *and* a central region created in the minds of people everywhere who have acted from a distance for its conservation, thus forming a globalized space or eco-borderland. Conservation and environmental concerns change local nature/culture relationships from afar and near. However, they also shape those far away, which challenges how we define borderlands and their interactions, encounters, social production, and remoteness. Journalist Richard Louv (2005) has addressed this environmental *remoteness versus closeness* challenge in his famous book *The Last Child in the Woods*:

> Today, kids are aware of the global threats to the environment—but their physical contact, their intimacy with nature, is fading. That's exactly the opposite of how it was when I was a child. As a boy, I was unaware that my woods were ecologically connected with any other forests. Nobody in the 1950s talked about acid rain or holes in the ozone layer or global warming. But I knew my woods and my fields; I knew every bend in the creek and dip in the beaten dirt paths. I wandered those woods even in my dreams. A kid today can likely tell you about the Amazon rain forest—but not about the last time he or she explored the woods in solitude or lay in a field listening to the wind and watching the clouds move.
>
> pp. 1–2

Therefore, we are encouraged to define ecological borderlands as fluid edges: fluid, not only implicating liquidness but including borderlands that are unstable, capricious, erratic, wavering, intermittent, and vague, and the humanly constructed social, cultural, and political concepts, such as borderlands. These eco-borderlands, as fluid edges, also entail addressing different scales: looking from where, and by whom, is a certain region an eco-borderland, accessible, distant, remote, or both?

According to our findings, Holmes (2016) has first used the term "eco-borderlands" in her book *Ecological Borderlands: Body, Nature and Spirit in Chicana Feminism*. Building on Mexican–USA borderlands and the Chicana literature (Anzaldúa, 1987) – a keystone in Borderlands Studies – she addressed the nexus of women and nature as well as *borderlands environmentalism*: on the one hand, a borderland marked by political paradigms, and on the other, a bioregion with transnational flows of people, flora, and fauna also subject to environmental activism. For Holmes (2016), borderlands environmentalism is tied to the environmental justice movement – particularly to issues related to land rights, water access, land use (animal grazing, *curandera* garden), displacement and evictions, workplace dangers (pesticides, chemicals), food insecurity, and healthcare access. Many of these intersectional issues, she has argued, remain ignored in the arena of environmentalism, which is thus ripe for critical borderlands analysis. Moreover, she has proposed new notions, such as ecological belonging in a translocal context, which emphasizes – as ecologists do in terms of bioregions – a sense of place and relationships between plants, animals, air, water, soil, and humans on the one hand, and – as environmental justice activists do in terms of intersectionality – on the other a sense of awareness of how social hierarchies evolve over time to give certain groups more access to land, water, and other resources.

Radding (2017) has also mentioned "Ecological Cultural Borderlands" with a focus on environmental history. Drawing on borders, frontiers, and borderlands, she discusses how natural environments have been integrated into her study of borderlands history – for example, how Indigenous people in the transboundary region of Sonora, Mexico, which is marked by aridity and degrees of nomadism, endured a colonial frontier linked to New Spain and the early formation of the Mexican nation-state. In this sense, for Radding (2017), borderlands ecologies are closely entwined with Indigeneity through mutual production of landscapes and people. Based on Ingold's (2000) interlinked qualities of nature and culture, she has defined ecological cultural borderlands as both physical spaces, based on experiences of their material reality, and cultural spaces of commingling and exchanges of interconnected quality.

Laako (2023) has suggested that eco-borderlands are fluid edges that allow us to immerse ourselves more deeply into the relationships of borderlands with nature: their edges, encounters, entanglements, crossroads, and borders. Eco-borderlands address borderlands as places of encounter between human populations or between culture and nature, not characterized exclusively by binary categories of clash or conquer/appropriation, but by how they dissolve into each other, marking continuity and creating symbiosis, or "contaminated diversity", based on collaborative encounters and survival, as illustrated by Lowenhaupt Tsing (2015):

> We are contaminated by our encounters; they change who we are as we make way for others. As contamination changes world-making projects, mutual worlds—and new directions—may emerge. (...) Everyone carries a history of contamination; purity is not an option. (...) This book argues that staying alive—for every species—requires livable collaborations. Collaboration means working across difference, which leads to contamination. Without collaborations, we all die. (...) Rather than seeing only the expansion-and-conquest strategies of relentless individuals, we must look for histories that develop through contamination. (...) Contamination makes diversity.
>
> pp. 27–29

Rather than hybridity, this form of thinking is based on mutual dependencies and symbiosis, which takes critical distance from the frontier concept of "the strongest one wins" or essentialist/purist viewpoints, whether related to natural environments or those of Indigenous cultures. Nonetheless, as frontiers, eco-borderlands address uncomfortable power relations by adding the ecological dimension. Instead of being limited to resource conquest or appropriation of natural resources typical for frontier approaches, eco-borderlands inquire: what kind of power relations, and *go-betweens*, are formed in the borderlands, and how are they shaped across different scales and time? How have eco-borderlands been shaped by societies or populations, and how have they shaped societies and populations in different times and at different scales?

Simultaneously, such an approach to eco-borderlands necessitates a critical analysis of ecological concepts: what is a forest and what are its edges? What is conservation and its borderlands? How are waterscapes created and what are their borders? Thus, eco-borderlands also implicate the exploration of borderlands of different interfaces, layers, and landscapes and other -scapes. Accordingly, eco-borderlands may expand (political, societal, ecological, and cultural) understandings of space, place, and territory, including different vertical and horizontal dimensions. Thus, eco-borderlands are middle grounds, go-betweens of peoples, human and species populations, inhabitants and livelihoods ecosystems, wildlife, flora and fauna, land-, water-, and forestscapes, as well as their interfaces. Eco-borderlands are fluid edges, which address multispecies worlds and encounters.

Figure 1.1 (Laako, 2023) synthesizes aspects that can be critically examined as eco-borderlands, focused on those fluid edges shaped by natural environments and concerns over their state.

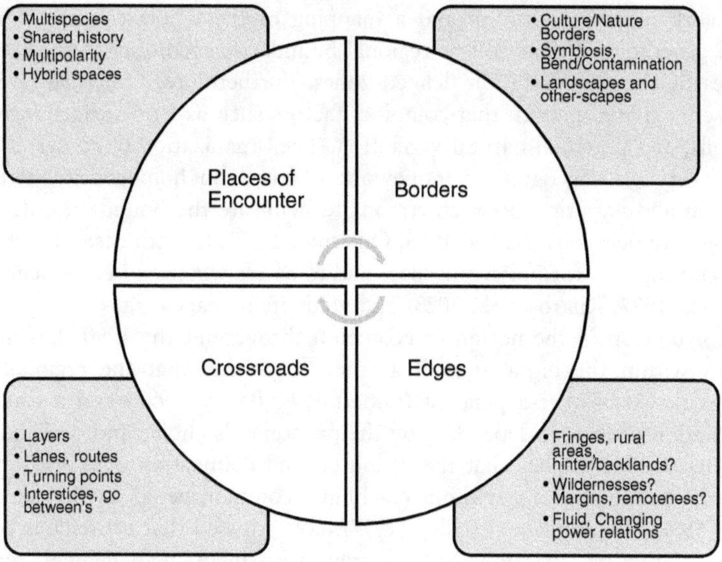

Figure 1.1 Conceptualization of Eco-Borderlands in the Case of the Maya Forest
Source: Laako, 2023, p. 80

From Ecoregions to Biodiversity Hotspots: Conservation at Borderlands

We have discussed the developments within the multidisciplinary Borderlands Studies as well as the ways to create bridges with environmental social sciences by means of eco-borderlands. In this section, we add the final piece to our concept of forest waterlands, which also addresses some related important ecological terminology. We begin by asking: *What can Borderlands Studies possibly have to do with Ecology and Conservation Biology?* We suggest that they might share more than would seem at first glance. As the above discussion has suggested, borderlands and eco-borderlands elucidate biogeographical areas and ecosystems defined as such by the previously mentioned fields of natural sciences. However, particularly with the development of Conservation Biology since the 1980s, the mapping and regionalization of certain areas with ecosystemic characteristics have increased, first for management and planning purposes, and later owing to the urgency of conservation. The most well-known example may be Amazonia. Depending on the historical context and the actors, these new ecosystem-motivated maps have been called *ecoregions or biodiversity hotspots* – both connected to landscape management and assessment, environmental geography, and, more recently, to biodiversity politics.

The term "ecoregion" appeared in the USA in the 1970s to classify regions according to their main ecosystems. Bailey (1983) has presented a first delineation of biogeographical zones based on climate and vegetation to obtain a four-level framework of regionalization and a mapping of the whole country. The delineated geographical units follow regional boundaries according to the ecological changes of climate-vegetation defined zones. Furthermore, Omernik (1987) has proposed a spatial pattern that combines factors such as land surface form, land use, soils, and potential natural vegetation. The organization of ecosystem information through a geographic framework is based on homogeneity within the ecoregion and variation between regions to delineate the boundaries, depending on subjective decisions (Bailey, 1983; Omernik, 1987). In both cases, the mapping has been proposed for policy purposes and to assess water and ecosystem quality (Omernik, 1987; Ruaro et al., 2023) and guide its management.

The evolution of the notion of ecoregion throughout the 1990s has included humans within the biota as well as the recognition that the regionalization framework is observer-dependent (Omernik & Bailey, 1997) and a result of a combination of factors depending on the proponent's choice and decision about boundaries. This implies that the definition and delineation of ecoregions vary according to the authors, without reaching a common perspective (Omernik & Bailey, 1997). Omernik and Bailey (1997) have stressed that ecoregions put forward a useful spatialization for ecosystem assessment, management, monitoring, and research, although they are not adequate for a sole resource and not interchangeable with terms such as watersheds or river basins, even if they can be complementary (see also Omernik & Griffith, 2014).

The concept of ecoregions has been exported to other parts of the planet (Ruaro et al., 2023) and is now used for biodiversity, conservation planning and assessment, ecosystem research, as well as protected-area management worldwide. Ecoregions are ecologically and geographically defined areas connected to landscape management and born out of the surge of interest in ecosystems (e.g., Bailey, 2014; Olson et al., 2001). Usually, they are defined by their homogeneity and distinctiveness from other ecoregions; however, as they do not have abrupt fringes or boundaries, they resemble borderlands as fluid edges. They include terrestrial, freshwater, and marine areas, which amplifies the understanding of the ecosystems, albeit also collocating them in differentiated units. Initially conceived as a pattern of ecosystem regionalization and a tool for analysis and management, ecoregions are today also used by some stakeholders to promote and identify priority areas for conservation (Olson et al., 2001).

The Maya Forest waterlands relate to different ecoregional delineations. The first refers to two ecoregional delineations proposed at the end of the 1990s. One was established by the Commission for Environmental Cooperation (CEC) in 1997 for Northern America and also exists in a more recent updated version (Wiken et al., 2011). For Central and South America, Griffith et al. (1998) have proposed a regionalization including the three-level-maps from previous proposals such as the biogeographical ecoregions used by the World Wild Fund (WWF) and The Nature Conservancy. Today, research on Mexican and Central American ecoregions refers to both (Ríos & Raga, 2018).

The second ecoregional delineation that partly coincides with the Maya Forest waterlands and is considered the first delineation of the "Maya Forest" is the "Maya Forest Ecoregional Plan" – an abbreviation of a longer name, the "Maya, Zoque and Olmec Forest Ecoregional Plan" (García & Secaira, 2006), based on cartography. The plan has proposed maps of a transboundary area of forests, wetlands, and savannas located between Mexico, Guatemala, and Belize, characterized by a large biodiversity but also by cultural diversity, in the past and at the time of its publication. Thus, the ecoregion's name refers to three cultural areas linked to their Indigenous populations. The ecoregional plan has focused on conservation through a network of conservation areas, regional alliances, and strategies. It has proposed a common agenda for conservation, established by national and international NGOs and research institutions, which has also been agreed with the national governments of Belize, Guatemala, and Mexico.

From the beginning of the twenty-first century, the creation of ecoregions has been connected to Conservation Biology, which differs from Ecology in that it focuses on those parts of nature that are in a state of flux, predicting the future and preserving biodiversity. Conversely, Ecology addresses the great patterns of biodiversity and their causes; in other words, as a discipline, it seeks to understand the processes that govern species' populations, communities, and ecosystems at or near equilibrium (Hintzen et al., 2019). According to Hintzen et al. (2019), Conservation Biology has always had a social-science part to it as "conservation is an argument among people". Some authors (e.g., Bennett et al., 2017) have even argued that environmental social science should be called "conservation social science". The objective of social sciences is to critically analyze power relations related to environmental issues, including conservationists (i.e., conservation biologists and other similar epistemic communities that implement policies in different locations); however, as observed, certain critical strands may also opt to support causes such as Indigenous people. The protection of vulnerable, marginalized people and regions is a central ethical criterion in social sciences, and the key requirement is to be free to critique power relations of all types. This has included social scientists themselves, which is another long debate that we cannot fully address here but is important to be mentioned. However, it is also worth noting that, particularly in Mesoamerica, conservation biologists, among others, have been active in integrating new, biocultural interdisciplinary research and *Biocultural Conservation*, which seeks to bridge biological and cultural diversities, although often focused on Indigenous people (e.g., Cocks, 2006; Laako et al., 2022; Stevens, 2014).

In addition to the creation of ecoregions, the Mesoamerican humid rainforest area – particularly the areas between Mexico, Guatemala, and Belize – has been identified as a biodiversity hotspot in 1999. Biodiversity hotspots are areas with high levels of endangered biological diversity, a notion which was first used at the end of the 1980s to describe areas rich in biodiversity and under threat (Myers, 1990). Myers first proposed ten tropical forest hotspots in 1988 and another eight two years later according to three longitudinal factors regarding plant species: their number at the origin, remaining in 1990, and likely to survive at the beginning of the twenty-first century (Myers, 1990).

Conservation International (CI) adopted Myer's proposal in 1989 and conducted a new assessment that included a redefinition of the term "biodiversity" hotspot in 1996. The number of hotspots was updated to 25 in 1999 and to 34 in 2005. Today, there are 36 global biodiversity hotspots, defined as a combination of at least 1,500 "endemic" plant species having lost at least 70% of their primary native vegetation (Critical Ecosystem Partnership Fund, 2024). Biodiversity hotspots are now conceived as biogeographic areas delineated by biological boundaries, based on specific biota or combinations of closed species (Myers et al., 2000).

The separation of Northern Mesoamerica from Southern Mesoamerica hotspots appeared in 2003 as a conservation strategy focusing on protected areas of the former. CI created an ecosystem profile that covered northwest Belize, north and central Guatemala, and the southern Mexican states of Campeche, Chiapas, Oaxaca, Quintana Roo, Tabasco, and Yucatán as part of this subregionalization. Actions were prioritized within two corridors, (1) the Selva Maya corridor and (2) the Selva Zoque and Chiapas/Guatemala Highlands corridor, which included eight key biodiversity areas (Critical Ecosystem Partnership Fund, 2004). The information updated in 2024 only mentions funding for six protected areas (Critical Ecosystem Partnership Fund, 2024). Biodiversity hotspots are constantly changing and being redefined owing to the transformation of threats and their impacts, and to new data collection, research (Mittermeier et al., 2004), and available funding.

Although different in their conception and origins, bioregional regionalization (ecoregions) and priority conservation areas (biodiversity hotspots) are now embedded in the conservation organizations' goals to assess conservation strategies. Therefore, the WWF has proposed the "Global 200" as exceptional biodiversity areas (Olson & Dinerstein 2002). These 238 ecoregions – 142 terrestrial, 53 freshwater, and 43 marine ecoregions – were selected based on common criteria: richness of species, endemism, exceptionality, rarity, and biological as well as conservation status.

Notwithstanding their apparent current complementary and overlapping elements, the Global 200 ecoregions and biodiversity hotspots do not coincide completely; all hotspots correspond to at least one of the 200 Global ecoregions, and approximately 60% of ecoregions include a hotspot (Mittermeier et al., 2004). Londoño-Murcía et al. (2010) has explained that the WWF 200 Global ecoregions only partially correspond to environmental and biological data on diversity and richness. Additionally, environmental diversity is not included as a factor in defining ecoregions, that is, as a main component of hotspots, and this represents limitations for conservation even for the 200 Global ecoregions. Thus, we can conclude that ecoregions and biodiversity hotspots, despite their complementarity and overlap, are different processes of regionalization due to their history and because their boundaries and components do not coincide (Londoño-Murcía et al., 2010). Consequently, Londoño-Murcía et al. (2010) have stated that protected areas fail to include some highly diverse ecoregions, which leads to the inadequacy of WWF ecoregions for conservation and management units and environmental representation – especially in regions with an urgent need for action, such as Mesoamerica.

Although our area of the Maya Forest is recognized as part of a broader Mesoamerican biodiversity hotspot, it does not appear as a sole ecoregion as such. The previously mentioned CEC (2023), for example, has identified the ecoregion "Tropical Wet Forest" of Soconusco, Chiapas, and another one limited to the Yucatán Peninsula's semi-evergreen tropical forest. The WWF Global 200 (2012), which resembles ecoregions, recognizes "Mesoamerican pine-oak forest" and "Tropical coral: Mesoamerican Barrier Reef System". In the WWF categorizations, Belize seems to be recognized in various sections, especially for its reef, freshwater and coastal corridors, and moist and pine forests extending toward Petén, Guatemala, and the Yucatán peninsula. For Mexico, the ecoregions of Yucatán's moist forest (Campeche, Quintana Roo, and Yucatán) are identified, in addition to the freshwater ecoregions of the Atlantic complex, Belizean lowlands, and Grijalva-Usumacinta. Guatemala shares some of these, particularly the Grijalva-Usumacinta ecoregion, Western Caribbean mangroves, and Yucatán/Atlantic moist forests and Central American pine oak forests. It is noteworthy that according to these ecoregion mappings, the Maya Forest contains many different, entangled marine, freshwater, and terrestrial ecosystems with a wide range of landscapes and other -scapes, which also extend beyond its supposed delimitations. This accurately illustrates the challenge of identifying the region with a single ecological term. A biodiversity hotspot may be the most descriptive; yet, these definitions also indicate the need to critically explore them, as well as other delineations such as protected areas and corridors.

Nevertheless, the tendency of the past decades to remap the world as biodiversity hotspots and ecoregions for conservation is interesting for Borderlands Studies for at least two reasons: (1) the mapping elucidates previously remote, peripheral borderlands now identified as biodiversity-rich, and brings them to global attention as new places of encounter, both between nature and culture and between people, including conservationist actors and those working in environmental management in general with the local residents, in addition to global audiences and activists impacting these regions from near and afar; (2) the conscious or unconscious labeling of borderlands based on natural environments. For example, the Great Plains are not only a borderland identified by Levin Rojo and Radding (2019), but also an ecoregion identified by the CEC in North America. Both borderlands and ecoregions exceed other human-made borders: they both take critical distance from political borders and integrate fluid edges and encounters rather than closed entities. Again, "Amazonia" and even the "Transatlantic Frame" may serve as examples of quite normalized obvious, and therefore invisible creations of eco-borderlands. We say "normalized, obvious and invisible" as the names of these ecoregions are often so much taken as given that we normalize them and stop thinking about who named them and on what grounds. Here, we refer to critical debates about such biogeographic definitions as "Latin America" versus "Abya Yala", or the latest literature discussing how Indigenous people influenced or were silenced about the knowledge of the natural world gathered by early scientists during colonization, which forms the building blocks of many disciplines of natural sciences (e.g., Del Pilar & Page, 2020; Pratt, 2008; Thurner & Cañizares-Esguerra, 2023).

Problematizing the Land–Water–Forest Nexus: Toward Forest Waterlands

The above-mentioned overlapping biogeographical and ecosystem delineations have been further used for conservation strategies and actions worldwide. In our region of study, this has also been the case, as we will see in the next chapters. In addition to different regional delimitations discussed by the academic literature and environmental policies and politics such as river basins, which will also be addressed in Chapter 4, the entanglements of land, water, and natural resources (forests in this case) that we propose to analyze occur in a peculiar context, namely, that of the borderlands. To understand the nexus in the land–water–forest triad, we propose the category of "forest waterlands," which requires considering the nexus between land and water as well as the links between land and forest. This is a key issue to define our proposed notion of forest waterlands.

Waters, lands, forests, and societies do not exist separately but are intertwined. The nexus between land, water, and forest that we propose to analyze through the notion of forest waterlands does not only refer to natural and ecosystemic links, but above all, it deals with the multiple entanglements between society and nature, which allows us to analyze the complex interactions between them. Some authors have conceptualized specific water–land interactions, such as in floodplains in diverse parts of the planet, which could be conceived as "human-water systems" (Di Baldassarre et al., 2013) or as "human-water-land systems" (Camargo, 2022). Our concept of forest waterlands is inspired by this new scholarship analyzing the land–water nexus as we consider the interconnections, entanglements, correspondence, and in-betweenness proposed by recent conceptualizations.

The first element that guides our discussion of the nexus between water and land focuses on the traditional separation between water and land in science and in Western thinking, which considers them as completely dissociable and in opposition to each other. Nevertheless, recent scholarship has contested such positions, and we will focus on this literature.

In the traditional social-science literature, water is considered a fluid and land a fixed entity. The dynamism of water is opposed to the immobility of the land, and they are conceptualized as different and disconnected; thus, a land–water dichotomy, opposition, or separation perspective (Camargo, 2022; Cortesi, 2016) characterized as indissoluble according to Lahari-Dutt (2014, p. 505) predominates. This conception is common in English dictionaries that define land in total opposition to water, and is also at the core of some disciplines such as geography (Lahari-Dutt, 2014). Geographers have traditionally tended to separate land and water, giving the land the privilege of having a higher status than water.

The literature also reflects this dichotomy. Swift's novel *Waterland* has clearly illustrated this Western perspective, which gives more value to the land and considers water uncivilized and a synonym of savageness. The idea of *terra firma* is central in Western philosophy and sciences. Controlling water to stabilize land as achieved by the British colonial power in Bengal (Lahari-Dutt, 2014) or freeing the land from water as described by Swift in *Waterland* (1983/2012):

> They ceased to be water people and became land people; they ceased to fish and fowl and became plumbers of the land. They joined in the destiny of the Fens, which was to strive not for but against water. For a century and a half, they dug, drained and pumped between the Bedford River and the great Ouse, boots perpetually mud-caked, ignorant of how their efforts were, little by little, changing the map of England. Or perhaps they did not cease to be water people. Perhaps they became amphibians. Because if you drain land you are intimately concerned with water; you have to know its ways. Perhaps at heart they always knew, in spite of their land-preserving efforts, that they belonged to the old, prehistoric flood.
>
> p. 13

Contrary to this traditional dichotomy, certain branches of the literature have emphasized the nexus between water and land instead of opposing them. In Latin America, Fals Borda (1979/2002), a Colombian sociologist, has proposed the concept of "amphibious culture", which he has defined as "a complex of behaviors, beliefs and practices related to the management of the natural environment, technology (productive forces) and the norms of agricultural production, fishing and hunting that prevail in the breeding communities of the Momposina depression".[1] The amphibious culture is therefore included among the manifestations of the superstructure of the society that inhabits this coastal subregion (Fals Borda, 1979/2002a, p. 21B).

The amphibious culture, a concept developed by Fals Borda but first proposed in the detailed and descriptive work of Luis Striffler (1886), a nineteenth-century French explorer of the numerous rivers of the Magdalena River basin, shapes a space of production and social reproduction based on water and land exploitation, with multi-labor activities based on water and land. From Striffler to more recent scholarship on amphibious culture, the amphibious population has been characterized by intertwined relations with water and land. Fals Borda (1979/2002a) has explained that "[t]hey seasonally combine agricultural, livestock, and forestry exploitation with river and fishery exploitation in the same habitat or territory" (p. 25B), following former descriptions (Striffler 1886, p. 80) and those subsequently followed by later authors (Altamar, 2021, p. 282; Mantilla, 2024, p. 72; McCrae, 2015, p. 81). This conception is completely opposed to the perspective of Swift's *Waterland*.

Some figures of the amphibious culture have emerged from the academic literature based on ethnography. In the Magdalena River basin, the *bogas* (boatmen) and local mythological references such as the turtle man (*hombre hicotea* in Spanish; Fals Borda, 1979/2002a, p. 27B) personify both the nature–culture intimate relation as well as the narrow relationship between water and land. In the third volume of his work, Fals Borda (1979/2002a) has transformed the omnipresent *hicotea man* into a symbolic characterization of the amphibious people – in other words, the local riparians. The *hicotea man* is then converted into a personification and anthropomorphic representation (McCrae, 2015), "a semi mythic persona" (Rappaport, 2020, p. 20) that characterizes the amphibious character and endurance. As a matter of fact, the *hicotea* turtle, *Emys decussata,* is characterized by its endurance and resistance to thirst and hunger.

The amphibious culture depends on adaptation to constantly changing local conditions (Fals Borda, 1979/2002a, p. 25B) and to climate and water movements – droughts and floods in their extreme manifestations, which shape a culture of resistance, endurance, and resilience (Altamar, 2021; Fals Borda, 1979/2002b; McCrae, 2015).

> Specifically, amphibian culture contains ideological elements and articulates psycho-social expressions, attitudes, prejudices, superstitions and legends that have to do with rivers, streams, ravines, slopes, sandbanks, marshes and rainforests; It includes institutions affected by the ecological structure and economic base of the tropics, such as linear settlement along watercourses, ways and means of natural resource exploitation, and some special patterns of land tenure.
>
> Fals Borda, 1979/2002a, p. 21B

Linear settlement refers to linear villages, with dwellings built along the riverbanks on dry land, surrounded by waters such as wetlands, marshes, and streams, which are typical in floodplains.

The term "amphibious" has been used as an adjective to refer to the water–land nexus in Colombia, but also, in other parts of the world, to analyze diverse related issues with floodplains, marshlands, and riverscapes. Gutiérrez and Escobar (2021) have proposed the "amphibious territory" (p. 75) to analyze the riverside, especially the surface beach composed by river sediments, as a space of interconnections between water and land, humans and no humans, and dispossession and conflicts. In her historical research, Mantilla Morales (2024, p.74) has referred to "amphibious landings", "lives", and "populations", explaining that the local riparians' skills are bound to water but also rooted in land. She has mentioned the concept of "amphibiousness" (p. 74) to outline constant interconnections between water and land, characterized as an inter-ecosystem symbiosis based on people being "equally knowledgeable about land and water … a culture equally at home in either space". Mantilla's definition of amphibiousness stressed a nexus that signifies giving the same value to both water and land and therefore to their diverse activities. In the European context, without citing previous Colombian references, Van Dam (2022) has proposed the concept of "amphibious culture" to explain how forms of adaptation to the wetlands in the Netherlands consisted in the compartmentalization of lands, the building of elevations for housing, and transport through waterways.

Parting from an "ever changing interplay of land and water", "amphibious anthropology" has been defined as

> A study of social and cultural life that takes into account its relations with the muddy ambivalences of delta environments, which vary between not quite open water, or sometimes one and sometimes the other, with water periodically in excess and repeatedly scarce.
>
> Krause, 2017, p. 1

Krause's (2017) analysis in river deltas is based on the four dimensions of hydrosociality, volatility, wetness, and rhythms. From a philosophical perspective, and following the German philosopher Sloterdijk, amphibious anthropology refers to "switching from one element to another" (ten Bos, 2009, p. 74), including water, land, and air. The confluence of land and water creates the production of space that is also related to time and movement (Gagné & Rasmussen, 2016). Both water and land are physical realities, but also the confluence of knowledge and abstract perspectives; this confluence defines the relational ontology that underlies an amphibious anthropology interested in this encounter of land and water (Gagné & Rasmussen, 2016).

In urban contexts, Ley (2021) has focused on amphibiousness as the results of a system of inequalities, which locates the urban amphibious as a figure of heterotopia, on "the border of a social (urban) order" (p. 59). The urban amphibious is the result of the relations between water and an infrastructure based on power relations, which confine certain categories of population to wetness. Ley (2021) has called for an amphibious anthropology that considers politics and governance in the context of climate change. In a similar context, Morita (2017) has referred to an "amphibious future" (p. 272) due to changes in a drainage basin caused by the water infrastructure. Nevertheless, research taking place in the Chao Phraya Delta in Thailand has presented the new conditions of amphibiousness as favorable.

Cortesi's (2016) research has taken place in floodplains. Moving away from the fact that rivers are considered as water only, owing to the ontological difference and separation posed between water and land by engineers, her fieldwork has highlighted the experiences of local stakeholders and ontology, which evoke a land–water nexus where "the river creates landscape". In North Bihar, India, Cortesi has observed a co-constitution of land and water through an ontological melting. Similarly, Camargo (2022) has addressed the water–land nexus by questioning the static characteristics of land (whereas water seems to be considered as dynamic), often based on economic expectation and presumption of stability necessary for property and taxes. Instead, he has maintained that land comes into being, expands, contracts, and changes under different temporalities as it interacts with other elements such as water. Camargo (2022) has provided an example from Colombian flood environments and the conflict among peasants, landowners, and the state for the appropriation of newly emerged fluvial land. Facing the traditional dichotomy and going beyond the separation, some authors have proposed the "wet theory" to evoke the water–land nexus (Appadurai & Breckenbridge, 2009). This theory analyzes places of "negotiations between land and sea" characterized by uncertainty and fluidity (Appadurai & Breckenbridge, 2009, p. viii).

Hybridity is another theoretical lens proposed by Lahari-Dutt (2017) to go beyond the water–land dichotomy and, furthermore, to stop identifying them as separate epistemic categories. The author has suggested that hybridity is a key element to understand wet theory. This "new water epistemology" includes land in a constantly changing environment conceived as a hybrid socio-nature that questions the idea of separation between nature and culture. Lahari-Dutt (2017) has also considered hybridity as a new methodology to approach specific landscapes such as soaked lands. In this context, Lahari-Dutt's (2014) notion of *waterlands* has reached beyond two dichotomies: the first one relates to the co-production of waterlands by both nature and culture, and the second one transcends the boundaries between water and land.

Parting from the clearly distinct categories of water and land, Cortesi (2022) has proposed the term "waterland" to elucidate how soil/land is not, in many cases, sufficiently separated from water to justify an oppositional category; these oppositional categories are often reinforced as separated blue and green units in biogeographical maps. Steinberg & Peters (2015) has eventually suggested "wet ontologies" to open the debate beyond fixity and to rethink boundaries. A suggestive vocabulary of interconnections between water and land has been proposed by these scholars of amphibiousness and wet theory. To illustrate the land–water nexus, Cortesi (2016) has evoked how "river creates landscape", "an ontological intimacy", "an intimate correspondence, tie, relation" or "a reciprocal cognitive correspondence". Appadurai and Beckenridge (2009) have dealt with "aqueous reality, uncertainties", a "spongy porosity".

Lahari-Dutt (2014, 2017) has suggested a vocabulary of empirical interconnections in association with the concept of water-land: "aqueous land", "muddy lands", "sponge environment", and "fluid landscapes" are examples of this epistemology (Lahari-Dutt, 2014). Thus, Cortesi (2016) has also explained that water and land transcend boundaries and are indissociable. Delta, floodplains, chars (lands that emerge from the riverbed in India), and sediment beaches appearing during the dry season in Latin America (Gutiérrez & Escobar, 2021) represent empirical examples and metaphors (Lahari-Dutt & Samanta, 2013) of hybridity, amphibiousness, and, of course, uncertainties. In the Bolivian highlands, Whitt (2022, p. 245) has analyzed the impacts of changes in "water–land cycles" – an alternance of droughts and floods – as a "water–land interface" during the year, conceptualized as a co-production of space that both depends on climate but also on politics. The ontological separation of water and land implies boundaries, lines of division, and edges. Appadurai and Breckenbridge (2009, p. viii) have considered that "edges are the historical product of a determined effort to imagine lines where none exist and then to make them survive in the face of an aqueous terrain which constantly defeats their materiality". Wet theory focuses on flux and crossing boundaries and considers in-betweenness at the core of the research. For Cortesi (2022), again, waterland is not simply "mud" but rather examples of invisible, hyporheic zones, such as rivers that include surface waters and sediments, ecology and hydro-eco-geology (interaction of water and soil underground), and watersheds as the basin that feeds it. Thus, for Cortesi (2022), waterland indicates an expanse, extension, and compound term created by joining two words for a completely new meaning, which may also involve rewriting the often-negative connotations of mud, stagnation, wetland, and swamp, illustrating entrapped flow and backwater that seem rotten or underdeveloped.

The forests are intertwined with the above discussion of *waterlands*, particularly in the traditional understanding of land as fixed and related to land-use, which involves the management, planning, and rearrangement of natural environments for human use, such as agriculture, pasture, and forestry schemes. Humans have, of course, modified and managed forests for millennia in diverse ways and for multiple purposes, including the formation of reserves. In this sense, forest conservation is not a new or uniquely modern phenomenon (e.g., Lehtinen, 2006). The forests, as ecosystems characterized by trees, also shift, change, and transition beyond human impact; thus, their borders are fluid. Forests regrow and appear where they were not before, while disappearing from where they should have been. Similarly, forests have been interpreted as endangered remnants, although later discovered as carefully managed by humans. This is, for example, the case of the Maya Forest, which has been rediscovered as a secondary forest rewilded during the colonial era (e.g., Laako et al., 2022; Miller, 2007; Poore, 2003). According to Côte et al. (2018), recent debates have addressed reforesting certain areas previously considered as forests but that were in the past regarded as natural grasslands. This pinpoints a conceptual struggle: where do forests start and where do they end? The answers, of course, vary according to who is using the term. While nowadays, these tend to be predominantly international organizations, nations-states, scientists, and conservationists, at the local level, forests might be identified by local people as something quite different (e.g., Cano, 2018; Côte et al., 2018). The names may also change according to the different stages and whereabouts of the forests.

Thus, as with *waterlands*, the definitions of forest are also sociopolitical and cultural, and they are not born in a vacuum (Chazdon et al., 2016). The difficulty, as Côte et al. (2018) have noted, is that the values embedded in such definitions are often silent and taken for granted until the forest boundaries are drawn; subsequently, they become visible, contested, and problematic.

The contemporary, predominant definition of forest is intimately tied to colonial history and, in particular, timber extraction. For example, Peluso and Vandergeest (e.g., 2011, 2020), in their many publications, have addressed *political forests* as part of the building of, at first, global forestry empires, followed by the postcolonial nation-states seeking to tame their "jungles" as part of the national territories and forestry schemes. In these schemes, forests came into being by being defined as such, separated from other land uses, such as agriculture or pasturelands. Simultaneously, forests were mapped and legislated. In Peluso and Vandergeest's view, the Food and Agriculture Organization of the

United Nations (FAO) replaced colonial empires as the key international network for political forestry. Currently, the predominant forest definition derives from the FAO (2020, p. 4) and reads as follows:

> Land spanning more than 0.5 hectares with trees higher than 5 meters and a canopy cover of more than 10 percent, or trees able to reach these thresholds in situ. It does not include land that is predominantly under agricultural or urban land use.

Indeed, the FAO defines forests as "wooded land". Given the component of "land", harvesting and clearing the area from trees is not considered "deforestation" when it is expected to regenerate naturally or with the aid of silvicultural measures (Chazdon et al., 2016). Hence, the definition is somewhat blind to whether the land is covered by planted or naturally grown forest, which has many important policy implications that exclude the consideration of an ecosystem or biodiversity, among others.

The contemporary history of forest definition derives from the 1700s timber management that predominantly considered that *forestlands* should be managed primarily to sustain timber production. This model of forestry had roots in a German-born scheme that regarded forests primarily as timber resources. The scheme then spread globally (Chazdon et al., 2016; Lehtinen, 2006). However, post-World War II, the FAO sought to conduct a global inventory of forest resources and adopted a new definition suitable for calculating wood harvesting potential. Simultaneously, it encouraged member states to adopt the same definition, which then became the most adopted forest definition globally. In 1953, the FAO defined forests as "land bearing vegetative associations dominated by trees of any size" (Chazdon et al., 2016, p. 541), whereas in the 1990s, this changed to the above definition, with emphasis on land with canopy cover.

Competing forest definitions have been proposed. One key paradigm emerges from conservation biologists and ecologists, who have questioned the FAO's definitions for "not seeing the forest for the trees" (Chazdon et al., 2016). They have maintained that forests are not only wooded lands composed of timber trees of certain coverage/boundaries, but ecosystems with fluid edges. The paradigm derives from the 1960s environmental movements that sought to employ an ecological definition of forests that would address conservation, deforestation, habitat loss, environmental degradation, and biodiversity decline. In this paradigm, forests became part of terrestrial ecoregions and bioregions. Thus, the conservationist definition of forest involves the understanding of intact forests that should be protected to conserve biological diversity; in other words, forest is defined as a complex of plant, animal, and micro-organism communities and their abiotic environment interacting as a functional unit, where trees are a key component (Chazdon et al., 2016, p. 541). As Côte et al. (2018, p. 254) have noted, the conservation-oriented forest definitions are also contested and pose theoretical problems related to tree–people relationships when entangled with policy and science categories such as "degraded", "untouched", and "primary/secondary".

In addition, Chazdon et al. (2016) have identified the forest definition categories used by groups focusing on climate change and earth stewardship. The former refer to forests as carbon sinks and emphasize reforestation to reduce climate-change effects. The latter refer to forests as complex systems involving both ecosystems and societies with an emphasis on ecosystem services. They also observe the emergence of a new paradigm focused on *landscapes*, which include forests and their surroundings as complex, adaptive systems composed of multiple eco-systems, which need to be managed. Thus, in this vision, forests are no longer isolated entities but integral components of dynamic, multifunctional landscapes. The new definition of landscapes blurs the boundaries of previous forest definitions employed by forestry, agriculture, and conservation institutions. Indeed, *political landscapes* that involve the complex interplay of disturbed forests, forest dwellers and people, forestry, and states have been subject to vigorous social scientific discussion (e.g., Boyer, 2015; Lounela et al., 2019; Lowenhaupt Tsing, 2015).

Evidently, the concept of landscape continues to be fixated on *land*. The land-fixed, terrestrial exclusivity of forests has been questioned at least by Wernberg and Filbee-Dexter (2019), who have argued that the FAO's definition of forests, in fact, can and should also be applied to "marine forests", which refer to seaweed beds.

Waters and forests are entangled and codependent in many ways. This is evident, in particular, in our case study of the Maya Forest, which is defined by the aim to conserve the humid tropical rainforest identified as the principal, connecting landscape of the region. "Rainforests", by definition, involve a watery element – particularly with reference to their shared characteristic of a moist environment with high average rainfall spread regularly over the entire year, thus with few dry periods (Smouts, 2003). The temperature is usually high but not excessive. The tropical rainforest was defined in opposition to dry, arid forests and especially in contrast to the grassy, sparsely treed savannas, and the term was apparently first used by the German botanist A. F. W. Schimper in 1898.

According to Poore (2003) and Corlett and Primack (2011), closed canopy tropical rainforests became widespread during the Paleocene (54–66 million years ago). Particularly interesting is the case of the Central American or Neotropical area where the Maya Forest is located, which generated its own particular type of tropical rainforest ecosystem with the formation of the land ridge three million years ago, which acted as a filter by both preventing some North American species from crossing to South America and allowing many others to pass through. Many South American species did not spread to North America, however (Corlett & Primack, 2011). Rainforests can be divided into lowland and montane ones, with the latter usually having their own, distinct ecosystem. In fact, Corlett and Primack (ibid.) have maintained that while it is easy to make generalizations about tropical rainforests, compelling as they are and thus easy to reduce to the single problem of "saving the rainforest", many rainforests succeed in a wide range of environments.

Tropical rainforests also tend to be the most species-rich of all terrestrial ecosystems – in other words, the combinations of water and forest elements feed the biodiversity and vice versa. Indeed, Smouts (2003) has underlined that as the rainforest soil is often poor, the forest, through its unique capacity to quickly recycle the nutrients through organisms such as leaves, insects, fungi, carcasses, and others in the humid environment, is what gives the rainforests their impression of abundance and exuberance. Forests regulate water drainage and protect against flooding; treetops hold rainwater, which evaporates into the atmosphere, and the roots absorb water from the soil and maintain watershed banks. Thus, Smouts (2003) has underlined that tropical rainforests are not just the "lungs" but importantly sustain water and climate cycles. Corlett and Primack (2011) have maintained that the high diversity of birds, plants, and mammals has been the most popular topic of tropical rainforest research, although in reality, such species only make up a tiny fraction of the total number. Indeed, they have pointed out that one particular group has been understudied in Tropical Ecology: insects. Most of the insect species in the tropics have not been explored (except for butterflies), although it has been calculated that there are over four million species of them. The ecology of tropical insects is largely unknown, even though whoever has been to a tropical rainforest has certainly made their acquaintance to some degree.

Tropical rainforests have been the subject of strong imaginaries of often somewhat insect-filled (albeit also romanticized) landscapes. The concept of the tropical rainforest has evolved from changing perceptions related to uninhabitable dark places, filled with diseases, to locations of conquest, violence, and conflict to lungs of the world, sacred places, and spaces for tourist adventures, to mention a few (e.g., Laako et al., *forthcoming*; Laako & Kauffer, 2022; Laako et al., 2022). Smouts (2003), who wrote an instructive book on the darker sides of global ecopolitics in terms of tropical forests and the International Tropical Timber Organization (ITTO), has highlighted that the state of the forests always reflects the circumstances at a given point in time. Tropical rainforests, together with the emerging concept of biodiversity, became a global political object of international negotiation in the 1970s, eventually leading to decisions and practices that affect a wide range of practices worldwide. The focus on tropical forests was new in the 1970s, when forests had been predominantly related to wetlands, aimed to protect waterfowl in mangrove and swamp forests, which importantly did not interfere with timber trade. Moreover, the relationship of ecologists with the tropical rainforests was dominated by the charismatic African species – birds, elephants, big monkeys – and great botanical gardens; later, it politicized the traditional livelihoods of local people that have now become illegal hunters and poachers involved in biodiversity smuggling (see also Vadrot, 2018). Shortly afterward, tropical forests emerged as heritage sites, although Smouts (2003, p. 47) has critically maintained that industrialized countries have also sought unrestricted access to the genetic resources found in the tropical forests of developing countries. These interests are then entangled with states that find their forest cover shrinking but intend to conceal this fact by constantly enlarging the notion of the forest (Smouts, 2003, p. 57).

These developments and paradigms related to forest conceptualizations and the paradigms behind them have led to such diplomatic definitions as the FAO's, which lacks rigor but allows some global measure (Smouts 2003). Thus, Smouts (2003) has argued that rather than to an eco-system, the term "tropical rainforest" refers to a certain type of social, political, and economic relation that structures the interactions of humans and societies with nature in a given moment, which are often marked by tensions and confusions. In the case of the tropical rainforest, the concept, in her viewpoint, has been useful in drawing attention to the alarming rate of deforestation and the problems encountered in these areas: the search for new farmland, the overharvesting of timber, illegal cutting-down of trees, unclear logging concession policies, and pillaging of Indigenous communities.

Thus, our term "forest waterlands" is derived from the analytical perspective of social sciences, in the case of the Maya Forest, and the need to expand analytically toward the viewpoint of the land–water–forest nexus with all its sociopolitical and cultural tensions. This is particularly important as we build on Borderlands Studies, but with a focus on the current tendencies of conservation and natural resource management, which rekindle the interest in the notion of eco-borderlands. This also means that our use of "forest waterlands", originating from the Maya Forest and Borderlands Studies, has different roots than that of Cortesi (2022), although we certainly share the critical approach in these oft-given categorizations, their fluid edges, and the attempt to explore beyond or to recognize their political boundaries. Forest waterlands are an intrinsically boundary and hybrid phenomenon, where land, water, and forest coincide and entangle in multiple ways. Our proposal of waterlands in the plural indicates that the water–land–forest nexus could be applied in more than one combination.

Forest waterlands are defined as borderlands, which are entangled, fluid-edged, and bordered regions located at the fringes of power centers and/or generally considered or constructed as remote, peripheral, and marginal spaces and landscapes. They are organized around the nexus of water–land–forest as common elements that define, delineate, and link but also separate spaces, which often have a history of exploitative and extractive natural resource management. Forest waterlands are also spaces of coming together and of crossroads based on waters, forests, and lands, although, in this book, we pay particular attention to the element of water as a hidden but unifying element.

Forest waterlands, based on borderlands, are also subject to ecological concerns as they are often perceived by conservationists as lands full of biodiversity. They are frequently located at international borders, although forest waterlands are particularly entangled and fluid. Along forest waterlands, waters are supposed to convert into borders to delineate diverse forms of boundaries and be the main sustenance of this type of borderland; however, water is fluid, and emerging boundaries are uncertain, undefined, and sometimes imaginary lines. Forest waterlands help to understand borderlands as fluid edges shaped by conservation, and as a result, as more "contaminated" spaces involving new natural/societal/political relationships and entanglements in the interstices yet characterized by important incongruences and discrepancies (social and political) that need to be outlined and exposed. This is why we also focus on politics and policies: conservation and environmental management operate through them, albeit contested within the borderlands in many ways.

Forest waterlands involve diverse sources of water (rivers, lakes, lagoons, lakes, caves, aquifers, and wetlands) as well as watered territories as spaces of connection (such as river basins) and empirical and socially, culturally, and politically constructed spaces that bound waters and biodiversity. Moreover, waterlands are defined as open spaces that link land to "larger waters", such as seas, through river continuity in some cases, but also through conservation projects and resource management that link waters and lands, and waterlands to maritime waters. Again, forests are linked to these as spaces for biodiversity conservation and as political forests defined by states with their land-use policies, which separate them from other uses such as agricultural zones. When located in borderlands, such forests become part of the management of interstate relations beyond domestic spheres. In this way, forest waterlands involve international relations and geopolitics, although this is often not recognized as such.

In Figure 1.2, we highlight the key aspects related to our analytical approach to forest waterlands, which integrates borderlands, with fluid edges of entanglements and in-betweenness, as well as the boundaries and discrepancies related to these often politically separated units. In the chapters that follow, we critically explore these aspects.

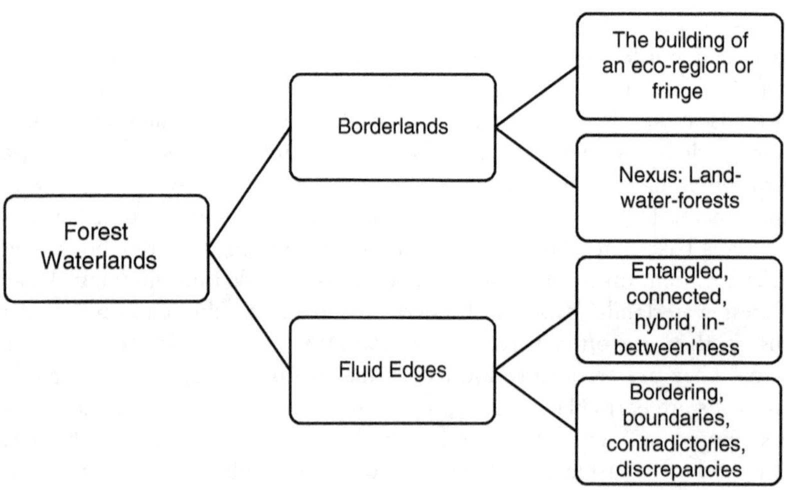

Figure 1.2 Conceptual Frame for Forest Waterlands

Conclusions: Forest Waterlands Are Political Borderlands

In this chapter, we have conducted a multidisciplinary exploration and analysis of developments in different fields that, in our viewpoint, contribute to the understanding of contemporary borderlands shaped by environmental concerns. While doing so, we also wanted to shed light on the multidisciplinary nature of Borderlands Studies and how it relates to environmental social sciences and Conservation Biology by analyzing regions – borderlands – often left underfoot of mainstream accounts. Yet, despite their shared interest in these regions, the fields have engaged little with each other to ponder about the current nature of cultural, political, and ecological borderlands. In this book, our objective is to explore these inquiries by means of a *forest waterlands* perspective that integrates the key aspects related to Borderlands Studies such as remote peripheries, crossroads, and places of encounters, while considering the creation of eco-borderlands and ecoregions, such as our case of the Maya Forest. At the same time, *forest waterlands* allow us to critically address the forest–water–land nexus, pondering about their entangled nature and separated political categories.

Indeed, as the above discussion implicitly shows, our book is focused on politics, given that conservation and environmental management are employed by political means in political spaces, while also creating politicizations and depoliticizations.

When dealing with conservation and environmental management, we also refer to *political borderlands*:

1. *Political* in the sense of entailing state intervention and policies, such as protected areas, water, environmental laws, and incentives
2. *Political* in the sense of entailing the intervention and influence of a political actor: conservationists and other environmental actors that seek to employ new measures for conservation and promote environmental priorities – in other words, a certain kind of relationship to natural environments and geographies
3. *Political* in the sense that other actors and inhabitants usually respond to the above-mentioned in different ways and pursue other paths, sometimes creating conflicts, sometimes collaboration, and often, as mentioned by Borderlands scholars, negotiation and *go-betweens*
4. *Political* in the sense of creating perceptions, definitions, even worldviews related to environmental issues, such as the opposing categories of water–land–forest, or a particular imaginary of the Amazonia as the "lungs of the world", the Antarctica as "empty lands", or the "tropical rainforest" as an exotic adventure, nobody's-land, hostile place, or biodiversity hotspot
5. *Political* because borderlands and the definitions of borderlands always depend on who is looking and from where – they are always *situated* – especially the very scholars of Borderlands Studies (Laako, 2016)

6 *Political* because the definitions of borderlands entail a political space or politicized lands and resources, and
7 *Political*, in our case of the Maya Forest, also because we deal with a transboundary region comprising international borders

Thus, borderlands are never only cultural and ecological; rather, they are political, even if we do not only focus on conflict and problems. This is nowhere as clear as in the following chapter, which zooms into the definitions of the Maya Forest, including the part missing from this chapter – *the Maya*.

Note

1 The Momposina depression around the town of Mompox is part of the Magdalena River basin, the biggest river and most important basin of Colombia. Three rivers flow and join at this fertile floodplain where daily life is punctuated by the confluence of water and land.

References

Altamar Lozano, J. P. (2021). *Análisis de la narrativa y el discurso en la literatura de viajes: Luis Striffler y el caribe granadino en el siglo* XIX. [Bachelor Thesis, Universidad Externado de Colombia]. Repositorio Universidad Externado de Colombia https://bdigital.uexternado.edu.co/bitstreams/0d53db68-5acc-48b1-aac4-fccd3e41c369/download.

Andersson, R., Cothran B. & Kekki, S. (2021). Traditional Indigenous Knowledge and Nature Protection. In Andersson R., Cothran B., Kekki S. (Eds) *Bridging Cultural Concepts of Nature: Indigenous People and Protected Spaces of Nature* (pp. 1–25). Helsinki University Press.

Anzaldúa, G. (1987). *Borderlands/La Frontera: The New Mestiza.* Aunt Lute Books, San Francisco.

Appadurai, A., & Breckenridge C.A. (2009). Wet Theory: a Foreword. In *Soak: Mumbai in an Estuary* (pp. viii–ix). Rupa & Co.

Bailey, R. G. (1983). Delineation of ecosystem regions. *Environmental management*, 7, 365–373.

Bailey, R. (2014). *Ecoregions: The Ecosystem Geography of the Oceans and Continents.* Springer.

Bennett, N.J., Roth, R., Klain, S.C., Chan, K.M.A., Clark, D.A., Cullman, G., Epstein, G., Nelson, M.P., Stedman, R., Teel, T.L., Thomas, R.E.W., Wyborn, C., Curran, D., Greenberg, A., Sandlos, J. & Veríssimo, D. (2017). Mainstreaming the social sciences in conservation. *Conservation Biology*, 31: 56–66.

Boyer, C. (2015). *Political Landscapes: Forests, Conservation and Community in Mexico.* Duke University Press.

Büscher, B. (2013). *Transforming the Frontier: Peace Parks and the Politics of Neoliberal Conservation in Southern Africa.* Duke University Press.

Camargo, A. (2022). Land born of water: Property, stasis, and motion in the floodplains of northern Colombia. *Geoforum*, 131, 223–231.

Cano, I. (2018). *De montaña a "reserva forestal": Colonización, sentido de comunidad y conservación en la selva Lacandona.* UNAM.

Carpenter, C. (2020). *Power in Conservation: Environmental Anthropology beyond Political Ecology*. Routledge.

Chazdon, R.L., Brancalion, P.H.S., Laestadius, L., Bennett A., Buckingham, K., Kumar, C., Moll-Rocek, J., Guimaraes, I. & Wilson, S. (2016). When is a forest a forest? Forest concepts and definitions in the era of forest and landscape restoration. *Ambio* 45, 538–550.

Cocks, M. (2006). Biocultural Diversity: Moving Beyond the Realm of 'Indigenous' and 'Local' People. *Human Ecology* 34, 185–200.

Commission for Environmental Cooperation (CEC): Terrestrial Ecoregions. (2023). http://www.cec.org/north-american-environmental-atlas/terrestrial-ecoregions-level-i/ (Consulted in September 2023).

Corlett, R. & Primack, R. (2011). *Tropical Rain Forests: An Ecological and Biogeographical Comparison.* Wiley-Blackwell.

Cortesi, L. (2016). An ontology of water and land in North Bihar, India. *Journal of the Royal Anthropological Institute*, 27(4), 870–889.

Cortesi, L. (2022). Hydrotopias and waterland. *Geoforum*, 131, 215–222.

Cortesi, L. & Camargo, A. (Eds, 2022). Land-Water Nexus Special Issue. *Geoforum* 131.

Côte, M., Wartmann, F., & Purves, R. (2018). Introduction: The trouble with forest: definitions, values and boundaries. *Geografica Helvetica*, 73, 253–260.

Crewe, R. (2017). Connecting the Indies: The Hispano-Asian Pacific World in Early Modern Global History. *Estudos Históricos* 30:60, 17–34.

Critical Ecosystem Partnership Fund. (2004). *Ecosystem Profile. Northern Region of The Mesoamerica biodiversity hotspot. Belize, Guatemala, Mexico.* Conservation International.

Critical Ecosystem Partnership Fund. (2024). *Biodiversity hotspot defined.* Conservation Internationalhttps://www.cepf.net/our-work/biodiversity-hotspots/hotspots-defined.

Del Pilar Blanco, M. & Page, J. (2020). *Geopolitics, Culture and the Scientific Imaginary in Latin America.* University of Florida Press.

Di Baldassarre, G., Kooy, M., Kemerink, J. S., & Brandimarte, L. (2013). Towards understanding the dynamic behaviour of floodplains as human-water systems. *Hydrology and Earth System Sciences*, 17(8), 3235–3244.

Djordje-Stojanovic, D. (2018). Post-modern Metamorphosis of Limological Discourses: From "Natural" Borders to Borderscapes, *Nagoya University Journal of Law and Politics* 276, 97–161.

Dudley, N. & Stolton, S. (2020). *Leaving Space for Nature: The Critical Role of Area-Based Conservation.* Routledge.

Duffy, R. (2022). *Security and Conservation: The Politics of the Illegal Wildlife Trade.* Yale University Press.

Emel, J. & Wolch, J. (Eds, 1998) *Animal Geographies: Place, Politics and Identity in Nature-Culture Borderlands.* Verso.

Fals Borda, O. (1979/2002a). *Historia doble de la Costa* 1. Mompox y Loba. Universidad Nacional de Colombia. Banco de la República. El Áncora.

Fals Borda, O. (1979/2002b). *Historia doble de la Costa* 3. *Resistencia en el San Jorge.* Universidad Nacional de Colombia. Banco de la República. El Áncora.

FAO (2020). Global Forest Resources Assessment 2020: Terms and Definitions. Working paper 188. https://www.fao.org/documents/card/en?details=ca9825en (Accessed in April 2024).

Gagné, K., & Rasmussen, M. B. (2016). Introduction–an amphibious anthropology: the production of place at the confluence of land and water. *Anthropologica*, 135–149. https://muse.jhu.edu/article/641045.

García, G. & Secaira, F. (Eds, 2006). *Una visión para el futuro: cartografía de las Selvas Mayas, Zoque y Olmeca.* Infoterra Editores, San José, Costa Rica, TNC.

Garfield, S. (2012). The Brazilian Amazon and the Transnational Environment 1940–1990. In Teoksessa Bsumek E., Kinkela D., Atwood M. (Eds) *Nation-States and Global Environment: New Approaches to International Environmental history* (pp. 228–251). Oxford University Press.

Guidotti-Hernández, N. (2016). Borderlands Scholarship for the Twenty-First Century. Review. *American Quarterly* 68:2, 487–498.

Gutiérrez Campo, R. D., & Escobar Jiménez, K. (2021). Territorio anfibio y despojo en una zona de humedales protegida del Caribe colombiano. *Revista de Estudios Sociales*, (76), 75–92.

Hämäläinen, P. & Truett, S. (2011). On borderlands. *Journal of American History* 98 (2), 338–361.

Hämäläinen, P. (2010). The Politics of Grass: European Expansion, Ecological Change, and Indigenous Power in the Southwest Borderlands. *The William and Mary Quarterly* 67 (2), 173–208.

Haraway, D. (2007). *When Species Meet.* University of Minnesota Press.

Hintzen, R., Papadopoulou, M., Mounce, R., Banks-Leite, C., Holt, R., Mills, M., Knight, A., Leroi, A. & Rosindell, J. (2019). Relationship between conservation biology and ecology shown through machine reading of 32,000 articles. *Conservation Biology* 34:3, 721–732.

Holmes, C. (2016). *Ecological Borderlands: Body, Nature, and Spirit in Chicana Feminism.* University of Illinois Press.

Ingold, T. (2000). *The Perception of the Environment: Essays on Livelihood, Dwelling and Skill.* Routledge.

Johansson, T. (2008) Beasts on Fields: Human-Wildlife Conflicts in Nature-Culture Borderlands. Doctoral Dissertation, Department of Geography, University of Helsinki. HansaPrint, Helsinki. <https://helda.helsinki.fi/bitstream/ handle/10138/21165/beastson.pdf?sequence=1&isAllowed=y>.

Krause, F. (2017). Towards an amphibious anthropology of delta life. *Human Ecology*, 45(3), 403–408.

Kullaa, R., Lahti, J. & Lakomäki, S. (Eds, 2022). *Kolonialismi Suomen rajaseuduilla.* Gaudeamus.

Laako H. (2016). Decolonizing Vision on Borderlands: The Mexican Southern Borderlands in Critical Review. *Globalizations* 13:2, 173–187.

Laako, H. (2023). From Wilderness Appropriations to Fluid Forest Edges: The Maya Forest Eco-Borderlands and –Frontiers. *Alue ja Ympäristö* 51:2, 65–85.

Laako, H. & E. Kauffer (2022). Between Colonizing Waters and Extracting Forest Fronts: Entangled Eco-frontiers in the Usumacinta River Basin. *Political Geography* 96, 102566.

Laako, H, & E. Kauffer. (2021). Conservation in the Frontier: Negotiating Ownerships of Nature at the Mexican Southern Border. *Journal of Latin American Geography* 20:3, 40–69

Laako, H., Ramos, D., Pliego, E. & Márquez, B. (2022). Eco-frontier building in the Maya Forest borderlands. *Estudios Fronterizos* 23, e105.

Lahiri-Dutt, K. (2014). Beyond the water-land binary in geography: Water/lands of Bengal re-visioning hybridity Kuntala Lahiri-Dut. *ACME: An International Journal for Critical Geographies*, 13(3), 505–529.

Lahiri-Dutt, K. (2017). An epistemological re-visioning of hybridity: Water/lands 1. In *Water, Knowledge and the Environment in Asia* (pp. 52–69). Routledge.

Lahiri-Dutt, K., & Samanta, G. (2013). *Dancing with the river: people and life on the chars of South Asia.* Yale University Press.

Lehtinen, A. (2006). *Postcolonialism, Multitude, and the Politics of Nature: On the Changing Geographies of the European North.* University Press of America.

Levin Rojo, D. & Radding, C. (2019). Introduction: Borderlands, A Working Definition. In Levin Rojo, D. & Radding, C (Eds). *The Oxford Handbook of Borderlands of the Iberian World.* Oxford Academic, 1–27.

Ley, L. (2021). The global swamp: the amphibious as a figure of heterotopia. In *Delta Life: exploring dynamic environments where rivers meet the Sea* (pp. 57–80). Berghahn.

Londoño-Murcía, M. C., Téllez-Valdés, O., & Sánchez-Cordero, V. (2010). Environmental heterogeneity of WWF ecoregions and implications for conservation in Neotropical biodiversity hotspots. *Environmental Conservation*, 37, 114–127.

Lounela, A., Berglund, E. & Kallinen, T. (Eds, 2019). *Dwelling in Political Landscapes: Contemporary Anthropological Perspectives.* Finnish Literature Society, Studia Fennica.

Louv, R. (2008). *The Last Child in the Wood: Saving Our Children from a Nature-Deficit Disorder.* Algonquin Books.

Lowenhaupt Tsing, A. (2015). *The Mushroom at the End of the World: On the Possibility of Life in Capitalist Ruins.* Princeton University Press.

Mantilla Morales, V. (2024). Amphibious landings: Free people of color, food supply, and contested land tenure on the Magdalena River network (1796–1806). *Atlantic Studies*, 21(1), 70–89.

McRae, D. (2015). El hombre hicotea y la ecología de los paisajes acuáticos en Resistencia en el San Jorge. In *Tabula Rasa*, (23), 79–103.

Metcalf, A. (2005). *Go-Betweens and the Colonization of Brazil 1500–1600.* University of Texas Press.

Miller, S. (2007). *An Environmental History of Latin America.* Cambridge University Press.

Mittermeier, R.A., Robles Gil, P., Hoffman, M., Pilgrim, J., Brooks, T., & Mittermeier, C.G. (2004). *Hotspots Revisited: Earth's Biologically Richest and Most Endangered Terrestrial Ecoregions.* CEMEX.

Morita, A. (2017). From gravitational machine to universal habitat: the drainage basin and amphibious futures in the Chao Phraya Delta, Thailand. *Engaging Science, Technology, and Society*, 3, 259–275.

Mustonen, T. (Ed, 2017). *Geography from the Margins.* Snowchange Cooperative. http://www.snowchange.org/pages/wp-content/uploads/2017/10/Ari_Lehtinen_engl.pdf (Accessed in April 2024).

Myers, N. (1990). The biodiversity challenge: expanded hot-spots analysis. *Environmentalist*, 10(4), 243–256.

Myers, N., Mittermeier, R. A., Mittermeier, C. G., Da Fonseca, G. A., & Kent, J. (2000). Biodiversity hotspots for conservation priorities. *Nature*, 403(6772), 853–858.

Neimark, B., Mahanty, S., Dressler, W. & Hicks, C. (2020). Not *Just* Participation: The Rise of the Eco-Precariat in the Green Economy. *Antipode* 52:2, 496–521.

Olson, D. M., & Dinerstein, E. (2002). The Global 200: Priority ecoregions for global conservation. *Annals of the Missouri Botanical garden*, 199–224.

Olson, D., Dinerstein, E., Wikramanayake, E., Burgess, N., Powell, G., Underwood, E., D'amico, J., Itoua, I., Strand, H., Morrison, J., Loucks, C., Allnutt, T., Ricketts, T., Kura, Y., Lamoreux, J., Wettengel, W., Hedao, P., Kassem, K. (2001). Terrestrial Ecoregions of the World: A New Map of Life on Earth: A new global map of terrestrial ecoregions provides an innovative tool for conserving biodiversity. *BioScience*, 51:11, 933–938.

Omernik, J. M. (1987). Ecoregions of the conterminous United States. *Annals of the Association of American geographers*, 77(1), 118–125.

Omernik, J. M., & Bailey, R. G. (1997). Distinguishing between watersheds and ecoregions 1. *JAWRA Journal of the American Water Resources Association*, 33(5), 935–949.

Omernik, J. M., & Griffith, G. E. (2014). Ecoregions of the conterminous United States: evolution of a hierarchical spatial framework. *Environmental management*, 54, 1249–1266.

Peluso, N. & Lund, C. (Eds, 2013). *New Frontiers of Land Control*. Routledge.

Peluso, N. & Vandergeest, P. (2011). Political Ecologies of War and Forests: Counterinsurgencies and the Making of National Natures, *Annals of the Association of American Geographers*, 101:3, 587–608.

Peluso, N.L. & Vandergeest, P. (2020). Writing Political Forests. *Antipode*, 52, 1083–1103.

Poore, D. (2003). *Changing Landscapes: The Development of the International Tropical Timber Organization and its Influence on Tropical Forest Management*. Routledge.

Prado, F. (2012). The Fringes of Empires: Recent Scholarship on Colonial Frontiers and Borderlands in Latin America. *History Compass* 10, 318–333.

Pratt, M.L. (2008). *Imperial Eyes: Travel Writing and Transculturation*. Routledge.

Radding, C. (2014). Environment, Territory and Landscape Changes in Northern Mexico during the Era of Independence. In Readman, P., Radding, C. & Bryant, C. (Eds). *Borderlands in World History 1700–1914* (pp. 65–82). Palgrave Macmillan.

Radding, C. (2017). Naturalizing Borderlands in Time and Space: Imperial Frontiers and Historical Indigeneities in the Americas. *Revista Habitus – Revista do Instituto Goiano de Pré-História e Antropologia*, 15:1, 5–20.

Ramutsindela, M. (Ed, 2014). *Cartographies of Nature: How Nature Conservation Animates Borders*. Cambridge Scholars Publishing.

Rappaport, J. (2020). *Cowards don't make history: Orlando Fals Borda and the origins of participatory action research*. Duke University Press.

Ray, S. & Maier, K. (Eds, 2017). *Critical Norths: Space, Nature, Theory*. University of Alaska Press.

Readman, P., Radding, C. & Bryant, C. (Eds, 2014). *Borderlands in World History 1700–1914*. Palgrave Macmillan.

Ríos, B., & Raga, G. B. (2018). Spatio-temporal distribution of burned areas by ecoregions in Mexico and Central America. *International Journal of Remote Sensing*, 39 (4), 949–970.

Ruaro, R., Gubiani, É. A., & Hughes, R. M. (2023). Omernik's Ecoregion Framework: a Legacy for Understanding Regional Patterns in Attainable Resource Quality. *Environmental Management*, 1–11.

Salisbury, D.S. & Fagan, C. (2013). Coca and conservation: cultivation, eradication, and trafficking in the Amazon borderlands. *GeoJournal*, 78, 41–60.

Smouts, M. C. (2003). *Tropical Forests, International Jungle: The Underside of Global Ecopolitics*. Palgrave Macmillan.

Steinberg, P., & Peters, K. (2015). Wet ontologies, fluid spaces: Giving depth to volume through oceanic thinking. *Environment and planning D: society and space*, 33(2), 247–264.

Stevens, S. (2014). Indigenous Peoples, Biocultural Diversity and Protected Areas. In Stevens, S. (Ed). *Indigenous Peoples, National Parks and Protected Areas: A New Paradigm Linking Conservation, Culture and Rights* (pp. 15–46). The University of Arizona Press.

Striffler, L. (1886). *El río San Jorge*. Corporación Autonóma de los Valles del Sinú y San Jorge-fundiación Universitaria Luis Amigo.

Swift, G. (2012). *Waterland*. Vintage (Original work published in 1983).

ten Bos, R. (2009). Towards an amphibious anthropology: water and Peter Sloterdijk. *Environment and Planning D: Society and Space*, 27(1), 73–86.

Thurner, M. & Cañizares-Esguerra, J. (Eds, 2023). *The Invention of Humboldt: On the Geopolitics of Knowledge*. Routledge.

Vadrot, A. (2018). Endangered species, biodiversity and the politics of conservation. In Kütting, G. & Herman, K. (Eds). *Global Environmental Politics: Concepts, Theories and Case Studies* (pp. 198–226). Routledge.

Van Dam, P. J. (2022). The amphibous culture along the Zuiderzee and the big rivers in the Netherlands, 1500–1850. In *A comparative reflection about the Yangtze River and the Rhine* (pp. 274–295). Zhong book Company.

Vannini, P. (Ed, 2024). *Mobilities in Remote Places*. Routledge.

Wakild, E. (2013). Environmental Justice, Environmentalism, and Environmental History in Twentieth-Century Latin America. *History Compass* 11163–176.

Wernberg, T. & Filbee-Dexter, K. (2019). Missing the marine forest for the trees. *Marine Ecology Progress Series* 612, 209–215.

Whitt, C. (2022). From flooded fields to a vanished lake: The politics of broken water cycles in the Bolivian Altiplano. *Geoforum*, 131, 243–251.

Wiken, E., Jiménez NavaF., & Griffith, G. (2011). *North American Terrestrial Ecoregions—Level* III. Commission for Environmental Cooperation, Canada.

WWF Global 200. (2012). https://www.worldwildlife.org/publications/global-200 (Consulted in September 2023).

Ybarra, M. (2018). *Green Wars: Conservation and Decolonization in the Maya Forest*. University of California Press.

Zimmerer, K. (2007). Cultural ecology (and political ecology) in the "environmental borderlands": Exploring the expanded connectivities within geography. *Progress in Human Geography* 31 (2) 227–244.

2 Borderlands, the *Maya*, and the Creation of the Maya Forest

Hanna Laako and Edith Kauffer

Introduction

> The Maya Forest is not just a forest or an ecology or ancient cities. It is a space of collaboration. A space that opens collaboration, specifically with the NGOs. It also provides a space: A space for funding, for discussions, for governmental awareness because there is movement… of NGOS, of communities… And it is not just a conservation issue but a social one. The Maya Forest has created a space beyond a geographical one: A one of communities, scientists, and conservationists.
>
> A conservationist in Belize City (Interview by Laako, June 2022)

As the above citation suggests, the Maya Forest is a concept born in the 1990s that refers to a shared transboundary conservation space (e.g., Laako et al. 2022). Prior and posterior to the 1990s, the region covered by the Maya Forest has been the subject of vigorous mapping of all kinds, including by the Indigenous people themselves (e.g., Toledo Maya Cultural Council & the Maya People of Southern Belize 1997; EU Delegation 2023) and, above all, of ecologists and conservationists. As in Chapter 1, some of these mappings have to do with the identification of eco- and bioregions, while others have focused on biodiversity hotspots with a particular emphasis on conservation (e.g., ECOSUR, 1995). They also overlap with each other and with many other mappings that have described the past decades of the Maya Forest. In line with Holmes' (2016) conceptualization of eco-borderlands discussed in the previous chapter, the Maya Forest can be characterized as one: as a borderland marked by political paradigms – such as Indigenous territories, political boundaries, and conflicts – and as a bioregion with transnational flows of people, flora, and fauna, which also enhances environmental activism.

The Maya Forest is an eco-borderland increasingly subject to ecological regionalization (creation of eco- and bioregions) and ecological belonging in a translocal context, as noted by Holmes (2016), by various actors, and, in particular, by the literature related to forest people, their relationship with nature, and the uneasy awareness of social hierarchies and the contested access to land, water, and other resources. Analyzing the Maya Forest as an eco-borderland brings forth the region as a discursive borderland in the interstices of various scientific disciplines, which helps to explain the creation of the concept (and its

DOI: 10.4324/9781003429050-3

links to the Mayas), and their respective working spaces in protected areas. At the same time, the analysis of the Maya Forest as an eco-borderland sheds light on what is included and excluded from the predominant Maya Forest landscape – for example, the contemporary struggles of the people that self-identify as Mayas and many other settlers and inhabitants who do not fit the existing categorizations.

To this end, this chapter also suggests how difficult it is to keep holding on to the traditional and predominant definition of the Maya Forest region as peripheral and remote borderlands in the light of the existing mappings, borderings, and megaprojects, which indicate multiple borderlands encounters, paradigms, and even hotspots. During its contemporary history, the region consisting of the Maya Forest has been predominantly described as peripheral – in other words, a borderland at the edges of empires (e.g., Laako 2016). This has also been a form of self-identification of the countries in question: the Southern Mexican borderlands have been considered a marginal, forgotten region in comparison with the national center, and the predominant political and academic focus on the Northern border with the United States (Kauffer, 2019). Indeed, the Mexican–US borderlands form one of the predominant loci and bases for Borderlands Studies. The case of Guatemala is similar in the sense that the Northern borderlands – the department of Petén within the Maya Forest – was considered a wasteland, empty land, or *baldíos* well until the 1980s (e.g., Grandia 2012). For the same reasons, both regions have also been subjected to modernization, colonization, and development projects. Belize, again, has often been peripheral among the three nations, given its smallness and distinct history as a British colony looking toward the Caribbean communities (Kauffer, 2023). The chapter challenges the assumptions often deriving from Borderlands Studies that "uninhabited" or sparsely populated territories cannot form borderlands. As this chapter shows in the case of the Maya Forest, many supposedly uninhabited and "empty" forest waterlands are in fact subject to borderlands encounters, contested, and politicized regardless of whether they are actually populated or not, such as the Columbia River Forest Reserve in Belize discussed below. The Maya Forest spaces are lived-in spaces and Indigenous territories, even when they appear uninhabited. At the same time, they are lived-in spaces where two or more hegemonies or groups of people entangle, which invites further curiosities over intertwined geo- and biopolitics.

This chapter aims to explore and discuss the ways in which the Maya Forest waterlands are actually *Maya*, and how this relates to borderlands as a critical paradigm. For this purpose, we first address the origins of the concept of the Maya Forest born in the interstices of the fields of Archaeology and Ecology. Both academic fields are orientated toward the same space: the Mesoamerican Forest containing both rich biodiversity to be protected and ancient Mayan ruins to be discovered and conserved (see Maps 0.1 and 0.4 in the Introduction).

While the Maya Forest, as a concept, is used to reflect on the Mayas and forests in the given space, it often overlooks *who* those that primarily work, produce, and reproduce the contents of this particular narrative are, and why they do so. This led us to identify different pillars of the (re)production of the Maya: (1) the Mayanism based on early explorations and giving origin to both scientific Archaeology studying the Mayas and the popularized version, often produced by Westerners, which also feeds tourism; (2) the governments and tourism industry, which build on the aforementioned Mayanism, particularly the latter; and (3) the Indigenous groups themselves, based sometimes on Mayan claims. The concept of the Maya Forest is a particular place where archaeological Mayanism and conservation, derived from scientific collaborations, come together with those with local livelihoods, which nowadays is often wrapped up in the expression *biocultural diversity,* yet excluded from many predominant Maya Forest narratives. However, the creation of the Maya Forest is also intertwined with governments and the tourism industry, which convert the Maya Forest into a different kind of a hotspot beyond the often peripheral seeming region. Finally, we allude to the self-identifying Mayas in their own borderlands, albeit now contained by political boundaries and borders of protected areas. These Mayas make many claims related to biodiversity conservation and Mayan ruins, as well as academic research, as illustrated in the polemics about bioprospecting and Indigenous rights addressed in the final sections of this chapter.

Hence, this chapter follows the structure of the three pillars of the (re)production of the Maya, first identifying the space in the interstices of Archaeology and Ecology at the origins of the Maya Forest-concept. This is discussed from two related angles: as an academic space based on the understanding of forest people, and a shared space of study of ruins and biodiversity, located in the borderlands of protected areas in the rainforest. We then proceed to discuss the second pillar of the (re)production of the Maya and its tense relationship with the Maya Forest: tourism and popular *mayanization.* According to Almeida Poot et al. (2022, p. 68), who build on the original work of Bastos et al. (2007), mayanization refers to the increasing number and diversification of actors, with or without Mayan origin, who identify as Mayas. It is also closely and increasingly related to the offer of products and services accompanied with the "Maya"-adjective as a commercial brand, especially within the tourism industry, where mayanization has often been equated with the *disneyfication* of the Mayan culture. Almeida Poot et al. (2022) argue that mayanization has implicated the creation of Mayan representations of the Mayas and Mayan culture that are socially elaborated both by Mayan and non-Mayan actors. These representations are then appropriated and re-signified by many other actors based on their imaginaries, beliefs, and ideologies that follow their own useful political, economic, cultural, spiritual, and even academic aspirations. Whereas Mayanism and Mayanists are especially related to Archaeology and their popular and academic extensions, mayanization is a broader tendency of self-identification and external appropriation currently linked to tourism in particular.

Finally, we address the Mayan voices of the Maya Forest as the third pillar: how is the Maya narrative presented by the people who identify as Mayas today, and how do they actually relate to the Maya Forest and forest waterlands? This latter part includes, in addition to an account of the contemporary Mayan struggles, two brief sections with examples, entitled "insights", which illustrate and describe these links in two different locations in the Maya Forest – first the Usumacinta River as a Mayan river in the conjuncture of archaeological sites and tourism, and second the Columbia River Forest Reserve as part of the contemporary Mayan movement in Belize.

Many conclusions can be drawn from the complexity of the issues examined in the chapter. First, the chapter addresses the Maya Forest as an eco-borderland, which, on the one hand, challenges the understanding of this region as peripheral, and on the other, suggests that borderlands, as discussed in the previous chapter, can encompass supposedly uninhabited regions that are both Indigenous and ecological hotspots. Second, the chapter shows how the concept of the Maya Forest emerged in the interstices of the specific scientific fields of Archaeology and Ecology, which comprise heritage sites and biosphere reserves – namely, the ruins and the biodiversity found in the surrounding forests – as their "working spaces" to be conserved. During the past decades, the two fields have been evolving in the same direction, toward the understanding of the Mayas as forest people, whose ancient ruins and forests need to be safeguarded. Simultaneously, the same Maya narrative is increasingly exploited by the tourist industry and reclaimed by certain groups who self-identify as Mayas to safeguard Indigenous rights. Sometimes, the latter two are in opposition to one another, which suggests that the Maya Forest is also a landscape of many struggles.

The Maya Forest People: An Eco-borderland in the Interstices of Archaeology and Ecology

In many ways, the Maya Forest has been a borderland *par excellence*. As discussed in Chapter 1, Borderlands Studies emerged as a critical paradigm that underwent a cultural turn, emphasizing Indigenous histories and livelihoods as well as entanglements between settlers, states, and imperial powers. Thus, the borderlands explore encounters between powers but also the continuity of Indigenous territories and livelihoods.

In the region currently defined as the Maya Forest, there has been one constant social and political actor from pre-Hispanic times until today: the Mayas. According to de Vos (2002, p. 62), the Mayas should therefore occupy a central place in the analysis of the region, as they constitute the only actors that have been present since the preclassic period that gave origin to the so-called Mayan civilization.

Who are the Mayas, the resilient and central actors of the Maya Forest? The origin of the Maya concept is, more than anything, a scientific construct. Escalona (2018) has shown that the Maya concept was first used by Stephens and Catherwood (1841) in the mid-nineteenth century. It was based on three components studied during their journeys all over the region. The first one was related to the ancient civilization of the *Mayapán*, located on the Yucatán Peninsula in Mexico. The second one referred to the ancient language studied and discovered at the archaeological sites. Finally, the third one was based on the contemporary language witnessed during their fieldwork (Stephens & Catherwood, 1841). As confirmed by Nigh (2002, p. 452), among many others, "until the final decades of the 20th century, none of the people we call the living Maya thought of their own identity in terms of this anthropological category". In fact, according to Restall and Solari (2020) the Mayas prior to the twentieth century never thought of themselves as such. It is important to note that many groups within the Maya Forest that may have been identified as having Mayan roots by outsiders do not identify as Mayas (e.g., Escalona, 2017). Instead, many groups refer to themselves based on linguistic references or, for example, by the names of the places of origin. Nevertheless, during the last decades, the self-identification as Mayas has been an important tendency in Indigenous movements and claims in Belize, Guatemala, and Mexico, with specific forms of organization and mobilization in each national context.

The connections between the archaeological sites and contemporary people in the Maya Forest, later placed under the banner of the "Mayas", were, however, created by explorers and researchers, particularly from the USA and Europe (Escalona, 2018). The wider application of the term began in the nineteenth century and in the context of the emergence of the contemporary academic field of Archaeology and the great scientific fascination with this ancient civilization. According to Escalona (2017), the creation of the Mayas is the result of interweaving colonial documents, objects – from ceramics to ruins – and spoken languages that he calls "the manufacturing of the Mayas" as a new epistemological framework in times of social and political transition (Escalona, 2017, p. 133). In this respect, Escalona (2018, p. 141) states that "the idea of a Mayan civilization is a product of contemporary world history, at the crossroads of political, symbolic and scientific processes".

The scholars of these Mayan studies are generally called (and call themselves) *Mayanists*, whose predominant interest has been the exploration of the 2,000 years of Mayan history prior to the contact period – in other words, the *Mayacene,* with reference to the *Anthropocene* defined by the human-impact on the Earth's ecosystems, which, in the former case, is that of the Mayas on their environment (Restall & Solari 2020). According to Wainwright (2008), Mayanism emerged in the nineteenth century after the discovery of the archaeological sites, which brought about the study of the Mayas by Europeans and North Americans. Mayanism also was part of the context of the nascent Museography in the US (Escalona, 2018). Thus, from the 1960s, the Mayas formed a category consolidated by North American research, particularly by disciplines such as Linguistics, Archaeology, and Anthropology (Escalona, 2018). It also resulted from an interest in object collections and the implementation of tourism and nationalism, especially in the case of Mexico (Escalona, 2018), in an international context of the formation of states.

Image 2.1 Mayan Ruins in the Calakmul Biosphere Reserve, Campeche, Mexico
Source: Hanna Laako, 2022

Although the fields of Archaeology and Anthropology are key in creating and reproducing the Maya narrative, especially since the nineteenth century, the Mayas have been constantly reinvented and rediscovered by the West since the Spanish conquest and colonization in 1502 (Restall & Solari, 2020), and also by various contemporary Indigenous groups, as we will discuss in the final part of this chapter. Nevertheless, to understand the creation of the Maya Forest, we need to consider the region as an archaeological space of Maya production and Mayanist reproduction. This suggests that the central actor of the Maya Forest has been a Mayanist creation, predominantly with reference to pre-Hispanic times, but also with close ties to its current context.

Regarding the understanding of the emergence of such a concept as the Maya Forest, it is important to understand not only the history of the people within the Maya Forest, but also the practice of Mayanism, which is still a well-funded study field, particularly in North America, and whose presence has also been central to the Maya Forest since the nineteenth century, when the archaeological exploration of the so-called Mayan civilization began. As mentioned by Restall and Solari (2020) and shown, for example, by our previous research on the eco-frontiers in the Usumacinta River Basin (Laako & Kauffer, 2022; see also Wainwright, 2008 and Escalona, 2017), both scientific and archaeological studies began in the nineteenth century with explorers who mostly traveled on rivers and published different kinds of writings about their adventures in the then-unknown Maya Forest territories and ruins (see Stephens & Catherwood,

1841; Stephens, 1843). During the early twentieth century, there was global archaeological competition for the discovery of "lost cities", ruins, and artifacts in the Maya Forest region, as narrated, for example, by Lamb and Lamb (1954) in the descriptions of their travels in the jungles of Chiapas and Guatemala at the end of the 1930s.

Since the early Mayanist studies, the field has evolved in many directions. Restall and Solari (2020) address two streams: one is the popularized stream that is – they argue – overwhelmingly amateur or nonacademic, speculative, and oriented toward outlandish explanations (lost continents, paleocontacts with aliens), but that nevertheless continues to reach wide audiences. The other stream is based on the premise that the ancient Mayan civilization is part of the larger Mesoamerican civilizations, which eventually led to the development of scientific Archaeology. Both narratives have an impact on the development of tourism in the region. The governments of Mexico, Guatemala, and Belize, together with the international tourism business, form one of the important pillars engaged in the contemporary creation of the concept of the Maya, which takes place in the interaction with the other pillars: the Mayanists and those self-defined as Indigenous Mayas.

The scientific debates and paradigms within the contemporary field of Archaeology are many and cannot be fully covered here. The centrality of the Mayanist frame, of course, does not imply that there is no empirical basis for such a category as the "Maya" (Nigh, 2002), and in this chapter, it is not our objective to argue about these empirical foundations. Rather, we show how the creation of the concept of the Maya has generated different narratives with broad implications for the contemporary territorial developments of the Maya Forest.

However, we briefly mention two persistent arguments that emerge from the contemporary Mayanist scholarship and are entangled with the Maya Forest. The first maintains that the classic Mayan civilization cannot be considered an empire or a unified state with centralized control; rather, that it comprises various groups speaking different Mayan languages related to each other. The Mayan territory was geographically divided by the mountainous areas of the highlands; the arid, almost riverless peninsula; and the tropical rainforest in the central parts. In other words, it was a *borderland* of sorts at the edges of other empires, constantly subject to encounters with other civilizations, whether through conquest and war or via trade relations. At the time of the contact period, some Mayan groups were formed in a dozen city-states, so-called *poleis*. Restall and Solari (2020, p. 63) suggest that the Mayas may have actually thrived – also ecologically – owing to the sociopolitical characteristics of sharing a common culture, yet without organizing themselves in a single political system. Instead, the Mayan territories were decentralized, small-scale, and often based on nomadic agriculture in diverse environments, which allowed them to flourish. For example, the archaeological sites of Piedras Negras and Yaxchilán, both located on the Usumacinta River riversides, were never part of the same political system, but permanently in conflict with each other (Obregón Rodríguez & Liendo Stuardo, 2016).

Second, the Mayanist scholarship has shown that the ancient Mayas did not collapse and disappear, but transformed and moved to other places. They have experienced transitions and, as a result, abandoned ancient cities or may have opted to live in their surroundings in the deep forest – in the borderlands of the *polis*. Some city-towns may have experienced conquest and decay, while others have thrived long after the supposed "collapse", such as the self-defined Mayas of the coasts and rivers on the eastern seaboard of what is today the coastal areas of the states of Yucatán, Quintana Roo, and Belize.

In many ways, the contemporary Mayanist scholarship seems to indicate that the Mayan territory had many of the characteristics of eco-borderlands: it was located at the edge of empires and also frequently populated at the edges of city-states with fluid, changing power relations with each other and within the landscape they inhabited.

In southeast Yucatán, independent Mayan polities existed until the Caste War in the 1840s, with the final subjugation of an independent Mayan polity by a Hispanic-led armed force in 1933 (Restall & Solari, 2020, p. 80). Nevertheless, the Mayan communities exist and thrive, and they have developed different strategies of resistance – among others, strategic mobility and migration. In fact, Restall and Solari (2020, p. 85) suggest the following:

> In many cases, population loss was the result of flight more than epidemic mortality. In all phases of the Maya–Spanish wars, from the sixteenth through nineteenth centuries, the Maya propensity to disappear infuriated non-Mayas. Some archaeologists have suggested that during the centuries of the Classic period, Maya families moved to escape oppressive regimes or instability caused by warfare or environmental change (they call this "the option of departure"). Tactical migration was thus likely a deep-rooted expression of resistance to demands by outsiders or local elites, a pattern of behavior that was exacerbated by the periodic invasion campaigns of the sixteenth century, which, in turn, helped ensure a pattern of archipelago colonization by Spaniards. Spanish settlement required sedentary Maya communities and was thus undermined when the latter became a moving target. The phenomenon intensified as the colonial period wore on, with tactical migration by Mayas in the seventeenth century causing the Spanish province of Yucatán to shrink and the kingdom of the Itza to grow.
>
> Restall and Solari, 2020, p. 85

These important findings support the recent archaeological and anthropological scholarship that has highlighted how many Indigenous or Native people in the Americas were actually more mobile and nomadic than it was at first understood by the Europeans and then written about by colonial historians. Among others, Graeber and Wengrow (2021) mention that the Indigenous lands often seemed unoccupied and empty to the European colonizers, owing to a different understanding of land ownership and the existence of mobile cultures, in contrast to the sedentary European land-ownership culture.

Mobility has also turned out to be challenging in the context of contemporary Indigenous rights, which need to be applied in a state-centric frame. One example is that of Belize, where in recent decades, the Mayan movement has been struggling to achieve state-level recognition of the land rights confirmed by the Caribbean Court of Justice. In this case, the Belizean government has attempted to question these rights based on the mobility and migration of Mayan people across the state-centric Guatemalan–Belizean ascendancy zone. Although we have found it important to mention nomadic cultures here in the context of the Mayas, we do not want to imply that the contemporary Mayas lack national pride or pursue "states within states". As shown by borderlands historians (e.g., Lakomäki 2015) and our interviews with some contemporary Mayan leaders in the context of this research, many Indigenous people have chosen to defend the existing nation-states and seek to be fully recognized as citizens of their respective countries.

Nevertheless, beyond the historical or contemporary mobilities and issues regarding the role of kingdoms and smaller communities that remain to be explored, we point to the tendency in recent archaeological scholarship to increasingly address the Mayas as forest people or forest garden people. In their brief introduction to the Mayas, Restall and Solari (2020) underline that contemporary Mayanist approaches have increasingly explored and questioned the ways in which the ancient Mayas survived, particularly at peak population points (with perhaps as many as 10 million people) in an environment that was and is challenging for humans, given the riverless scrub forest, tropical rainforest, and mountain areas. The current Mayanist premise tends to be that the Mayas adapted to the lands and forests in many ways while maintaining diversity, from the more predominant crop/maize farming to forest gardens, from beekeeping and fruit tree cultivation to the hunting and domestication of animals. The current understanding is that farmers sought to keep the rainforests mostly intact by avoiding overcollecting, overfarming, and monocultivation.

As maintained by Ford and Nigh (2015, pp. 22–24), the assumption of a Mayan collapse due to deforestation is questionable. Their finding is that the early Mayan settlements depended on horticulture and an expansion of agriculture, while *living in the forest* (our emphasis) with the *milpa* system, defined as a resource-management system or a system of agroforestry which shaped the Maya Forest to meet subsistence and tribute needs, develop the political economy, and promote local and long-distance trade. The *milpa* forest garden system cycles over decades and suggests mobility. Indeed, Ford and Nigh (2015) maintain that the long-lasting cultural continuity of the Mayas in itself must be an indication of a human–environment relationship in which they lived holistically in the forest. The existence of an eco-borderland of sorts is based on fluid edges between human activities and nature. The understanding of the Maya Forest as the people living in the forest with their agroforestry system has transformed the forest from a supposedly pristine one into a forest co-created over a long time by its inhabitants; in other words, the Maya Forest has been shaped by both the agroecological *milpa* system and the forest garden, which, as pointed out by Ford and Nigh (2015, p. 13), disturbed the natural environment, but also worked with the forest and are integral to its creation and sustainability. For Ford and Nigh (2015), Maya Forest people are farmers who bring together *milpa* agroforestry and forest gardens with the forest – in other words, a livelihood and conviviality of people and forests by agricultural means. The Maya Forest is thus the result of prehistoric, colonial, and recent human activities, "[a]nd accomplished traditional farmers are creatively thriving on the forested landscape that some archaeologists have called a "green hell" (Ford & Nigh, 2015, p. 15).

The "green hell", however, is not only the space for archaeologists working to obtain findings related to the ancient forest people, but increasingly, a conserved forest considered ripe for biodiversity politics – in other words, an eco-borderlands for a different type of human-nature encounter: that of conservation linked to forest people, heritage, and biospheres (e.g. Fisher & Chase 2021; Juniper 2018). The Maya Forest concept describes a particular space formed in the interstices between the fields of Archaeology and Ecology in the "green hell", surrounded by the Mayan ruins and tropical biodiversity.

Forest people is a notion that began to be employed by the Food and Agriculture Organization of the United Nations (FAO) in the 1970s, although it seems to have first been used by Turnbull (1961) in his book *The Forest People*, an ethnography of Mbuti pygmies in Congo. The FAO's notion of *forest people* predominantly addressed rural forestry for local community development (Arts et al., 2012). The idea was to draw attention to the ways in which

Image 2.2 A Protected Forest
Source: Hanna Laako, 2023

the role of forests could meet the people's livelihood needs, implicating the emergence of a new approach to social forestry with increased community participation in forest resource management. According to Arts et al. (2012), in the same period, the term "agroforestry" was coined by the Canadian forester John Bene. Agroforestry, as described by Ford and Nigh (2015) when writing about the Mayas as *milpa* forest garden people, seeks to "integrate trees, food crops and/or animals in a combined production system compatible with the cultural practices of the local population" (Arts et al., 2012, p. 15). Indeed, most scholarship currently using the term "forest people" is related to agroforestry and Indigenous rights, and the NGOs working on the same notion such as the Forest People Programme connected to Survival International and the World Rainforest Movement have similar aims (e.g., Stevens, 2014; Colchester, 2003). They have all addressed and intended to solve the earlier conflict between the aspirations of Indigenous people and those of conservation planners, which emerged in particular with the issue of livelihoods and slash-and-burn methods in the 1990s. A prime example of this sort of agroforestry and forest people in collaboration with conservation is the Maya Biosphere Reserve's multiple-use zone of forest concessionaries, which is also discussed in the Introduction to this book and in Chapter 5.

Thus, in parallel – albeit as a separate development – a paradigm shift occurred in conservation circles linking conservation and development, which gave rise to several community-based conservation approaches such as transition zone management, integrated conservation and development projects, and community conservation, later facilitating community-based natural resource management with a focus on human-nature coexistence (Stevens, 2014; Colchester, 2003; Juniper, 2018). In the 1990s, this paradigm gained in strength with the idea that communities are the best caretakers of their proximate resources and enhanced collaboration with grassroots movements, Indigenous people, and NGOs. The strand has sought to combine the goals of biodiversity conservation and rural development. Simultaneously, a rights-based approach to conservation was developed by the International Union for the Conservation of Nature (IUCN) and the Parties to the Convention on Biological Diversity in the 1990s (Stevens, 2014). Prior to the 1990s, there were few of these sorts of considerations. The rights-based approach is substantial in acknowledging the problem of internal colonialism as a contemporary struggle of Indigenous people that often clashes with state-led conservation aims within the same, biodiversity-rich, and Indigenous territories. Although there have been instances when rights-based conservation has been successful, which need to be recognized (e.g., Andersson et al., 2021), many challenges continue to exist.

Later, the rights-based conservation approach evolved toward the notion of *biocultural diversity*; this aims to illustrate the link between cultural and biological diversity, which are considered interdependent and even co-evolved (e.g., Stevens, 2014). According to Stevens (2014), biocultural diversity originated in the works and mappings of Mac Chapin in Central America at the beginning of the 1990s, when it was found that there is an overlap between Indigenous people' territories and the remaining areas of high biodiversity and regions with relatively intact ecology; in other words, the Indigenous people tend to live precisely in the ecosystems that the conservationists are eager to protect. Indigenous territories, identified as borderlands, have thus become *eco-borderlands*, namely, spaces for conservation and biocultural diversity. Simultaneously, they are powerful places of encounter between new development paradigms, conservation, and livelihoods in spaces of biodiversity abundance, which may have seemed to be empty spaces for many Borderlands scholars.

Thus, the parallel developments in the two scientific disciplines – Archaeology and Ecology – have drifted toward the creation of a particular space and encounter of both archaeological tendencies, moving toward the understanding of the Mayas as forest people, and the conservation tendencies increasingly leaning toward rights-based biocultural diversity protection, as the foundations of the Maya Forest concept. This particular interface between Mayanists and conservationists is both physical – taking place in specific protected areas in the "green hell" – and theoretical, as it has given rise to a new paradigm, expressed in the "Maya Forest". It has led to expanding scholarship with this title (e.g., Nations, 2006; Primack et al., 1998; Ybarra, 2018) and a broad web of different kinds of protected areas and heritage sites. However, as shown in the introduction to this book, there are also controversies, such as the case of the Mirador-Río Azul basin campaign.

Image 2.3 Ongoing Archaeological Work in a Protected Area in the Maya Forest
Source: Hanna Laako, 2023

The Maya Forest as a geographical space and a concept created in the interstices of Archaeology and Ecology comprises all kinds of protected areas, particularly international categories such as biosphere reserves, Ramsar wetlands (albeit less discussed in the Maya Forest literature so far), and heritage sites, which are also forest waterlands intertwined with the Mayan narratives and archaeological ruins. These were considered the "Arch" of the Maya Forest by the early Maya Forest scholars (e.g., Primack et al., 1998), which is no coincidence. First, the UNESCO biosphere reserves and heritage sites in particular emphasize education and research, thus enabling the continuing work of archaeologists and ecologists. Second, these are international programs that promote visibility, funding, and the generation of different revenues such as tourism. They *internationalize* the given protected areas – in other words, they open up these areas to more global connectivity and collaboration, even transboundary conservation. Hence, they may introduce a system of checks and balances in these protected areas: on the one hand, they may allow establishing conservation aims in areas of existing livelihoods; on the other hand, they may foster nature and heritage conservation in areas where the governments would otherwise not be able to implement robust conservation strategies.

The heritage sites are based on the 1972 UNESCO World Heritage Convention, which identified the need to protect monuments, groups of buildings, and sites with universal value from the viewpoints of history, art, and science, as well as natural features, geological, and physiographical formations and natural sites with universal value from the viewpoints of science, conservation, and natural beauty (World Heritage Convention, 1972). Biosphere reserves are based on the 1971 UNESCO Man and Biosphere Programme, an intergovernmental scientific program to enhance the relationship between people and their environments. It "combines the natural and social sciences with a view to improving human livelihoods and safeguarding natural and managed ecosystems" (UNESCO MAB website, 2024). Thus, biosphere reserves are a sort of laboratory for sustainable development rather than exclusively protected areas, although they are categorized as such in some countries as for example Mexico. The Ramsar sites are based on the 1971 Convention on Wetlands, an intergovernmental treaty that provides a framework for the conservation and wise use of wetlands and their resources (Ramsar website, 2024). Laako et al. (2022) identified, among others, the following UNESCO-recognized protected areas in the Maya Forest borderlands: *Calakmul* (both a biosphere reserve and heritage site in Campeche, Mexico); *Montes Azules* (biosphere reserve in Chiapas, Mexico), *Pantanos de Centla* (biosphere reserve in Tabasco, Mexico); *Sian Ka'an* (both a biosphere reserve and a heritage site in Quintana Roo, Mexico); *Tikal* (heritage site in Petén, Guatemala); the *Maya* and the *Chiquibul Maya Mountains* (biosphere reserves in Petén, Guatemala), and the *Belize Barrier Reef System* (heritage site in Belize). These are also shown in Map 0.1 in the Introduction, where the main protected areas of the Maya Forest are described.

While Belize currently has no biosphere reserves, this model was also considered for Belize in the 1990s to foster the links between Archaeology and Ecology, with emphasis on the Mayas and forests, and to enhance the importance of the Chiquibul National Park and the Caracol Archaeological Site as tourist attractions, which form part of the Maya Route developments:

> Recent estimates indicate that 80% of the 1988 tourism came for forest related reasons: natural history (ecotourism) and archaeology. It is well established that archaeological based tourism can not be separated from ecotourism; without forests it is highly unlikely Maya ruins would attract as many people. Although Tikal is the major archaeological destination for tourists in Guatemala, the majority are ecotourists who come specifically to experience its tropical forests. It is no secret that a great many of these ecotourists actually travel through Belize to reach Tikal. A Biosphere Reserve with the subsequent development of tourist destinations will assist in capturing a larger market share of these tourist dollars.
>
> Miller & Miller (WCS), 1990, p. 6

Indeed, in the following section, the chapter addresses the links between tourism and the Maya Forest with a focus on the second pillar of the (re)production of the Maya.

The Uneasy, Entangled Maya Forest Routes: States and Tourism

> What is the Maya Forest? A cultural and natural heritage of archaeological sites? A strategy of conservation, a space to implement projects. It does not reflect local perspectives. And then there is the "light version" of the Maya Forest: A heritage or landscape with a social and economic logic to generate income and an economic motor… (…) The controversies are not discussed. The contexts are different, and there is no common recipe. The landscapes and geopolitical contexts are very different.
>
> A conservation leader in the Maya Biosphere Reserve (Interview by Laako, June 2022)

While the region of the Maya Forest may be a borderland – and an eco-borderland – *par excellence*, its contemporary nature as a periphery can be subject to critical examination. The second pillar of the (re)production of the Maya, that of tourism, has been so successful that it is worthwhile to ponder how marginal for example the Yucatán peninsula is, for being one of the most central tourist destinations in the world. Different Mexican government websites and newspapers (such as the *Cancun Sun*) have calculated around 20 million tourists yearly passing through the Cancun airport in Quintana Roo, and the numbers are growing after the COVID years. Most tourists come from North America and Europe, although the numbers nationally and from other Latin American countries are also increasing. For example, the new Tulum international airport was built on the Yucatán peninsula and inaugurated in December 2023 to promote tourism in Riviera Maya, where the Mayan narratives of archaeological ruins and *cenotes* (sinkholes), together with the sandy beaches by the Caribbean Sea, make for an attractive package. Tulum was formerly a small fishing village located south of Cancún, but has now been converted into an iconic touristic place, offering the full Mayan tour combining sandy beaches Mayan history and ruins, and a world full of *cenotes*, previously hidden in the forest community waterlands and situated at the doors of the Sian Ka'an Biosphere Reserve. Tulum's recent history in the touristic boom may be considered a typical case of the Caribbean Karst Waterlands, which combine the *Maya*, the forest, groundwaters, and pressures on lands facing touristic expansion (Torres-Mazuera, 2022).

As mentioned in the previous sections, two streams can be identified in Mayanism: one that took the direction of scientific Archaeology, and a popularized one. It could be easily argued that the popularized stream is the main one also feeding the pillar of tourism with the (re)production of the

Image 2.4 A Wall Painting in San Ignacio, Belize
Source: Hanna Laako, 2023

Maya. Indeed, in the Maya Forest region – especially in Riviera Maya – it is not too difficult to evidence this reproduction in the names of the hotels, gas stations, shops and bakeries, live shows, amusement parks, and anything that can be imagined to attract tourists, as part of the Mayan narrative built for tourism. The Mayas are currently part of the artificial Riviera Maya-landscape and narratives, some out of necessity to survive and others by choosing to adapt (e.g., Almeida Poot et al., 2022). Consequently, the diverse Mayanisms are entangled rather than differentiated and co-constructed. According to Pieck and Moog (2009), entanglements illustrate shifting configurations of actors and institutions within and across different sectors, implicating both mutual collaborations and contentious interaction and flows. Over time, some key entanglements may constitute spaces where influential discourses enfold, new projects are initiated, and material resources flow (Pieck & Moog, 2009).

A similar development also happens to the rainforest: the rainforest and Mayan ruins are both ripe for tourism. Among others, Nations (2006, preface) introduced and elaborated the concept of the Maya Forest by narrating a "unified story" of the region and including some "quick tours" and "an explanation of how the expanding industry of eco-tourism helps protect both national parks and archaeological sites". As indicated in the previous section, the value of lush forests surrounding the ruins and other attractions for tourists has been understood for a long time (Miller & Miller, 1990). In the case of the Maya Forest waterlands, water largely contributes to the touristic industry with surface waters – rivers, lakes, and waterfalls – and undergrounds waters (waterholes); however, in this case, it is also combined with the sea and beaches.

The fascination with tropical paradises and wild jungles dates back to early explorers and scientists similar to the Mayanists; often these two categories go hand in hand. Their writings have inspired others, and with modern, affordable transport and advertising, the "last great rainforests" have become accessible to large numbers of tourists. Scholars have often cited the role of the *National Geographic* in promoting tourists visiting the rainforest by introducing the Maya Route (e.g., Nations, 2006; Fedick, 2003). According to Fedick (2003), the *National Geographic* published 31 articles between 1913 and 2000 that included descriptions of the forest environment of the Maya Lowlands and also, particularly from 1985 onward, its destruction. In 1989, *La Ruta Maya* (the Maya Route) was introduced in an editorial on the protection of the tropical rainforest. The Maya Route was identified as a multinational plan to promote tourism in the Maya region. Fedick (2003) argues that approximately from 1975 onward, "rain forest" became the more predominant term for the "paradise" formerly called "jungle". The shift in the perceptions of the rainforest is well described by Slater (ed., 2003), who shows how the discourse and imaginary of rainforests have been constructed in tourism as "icons" and "spectacles". As these jungles were later rewritten by conservationists as lands full of biodiversity, a paradox occurs. This involves, on the one hand, the need for the remaining rainforests to be conserved, implicating simultaneously the need to convince governments and the public for their appreciation; on the other, their escalating interest to visit and exploit them in the form of icons and spectacles, interchangeably called the *mayanization* or d*isneyfication* of the Mayan culture and the Maya Forest.

The ways in which the three nation-states of the Maya Forest – Belize, Guatemala, and particularly Mexico – have generated such key spaces for Mayan rainforest tourism is remarkable, although not particularly conservationist. Yet, it must be pointed out that both scientific archaeologists and conservationists of the Maya Forest have contributed – voluntarily or involuntarily – to the development of tourism in the region. For example, to fund and obtain research permits for the Maya Forest, researchers and conservationists need to justify the importance of their explorations to governments and funders. Often, this is done by emphasizing cultural and natural heritage – the Mayan specificity – and its wider attraction.

The same applies to protected areas: with scant and shaky national funding for conservation, the project's applications and revenues tend to be justified with tourist attraction. A conservation leader in Laako's interview in June 2022 in Petén emphasized that their long experience with the governments is characterized by the constant entanglement of two different logics: their own logic of conservation and the governmental logic of revenue. In his view, discussions with governments always result in the same administrative inquiry: *what revenues will be generated by this territory, this protected area – what is the economic contribution provided by this landscape?* Thus, his organization has been constantly drafting economic strategies destined to convince the governments about possible revenues from conservation and the protected areas. Eco-tourism has indeed been the preferred option: responsible, ecologically conscious, and selected tourists that help to reduce – not to augment – the environmental impact. Of course, this is easier said than done.

In 2019, another conservationist and government representative interviewed by Laako in Chetumal, Mexico, told the history of the formation of the Maya Forest concept as a result of negotiations between conservationists preoccupied with the threat of losing the largest forest in Latin America after the Amazon, which had to be named in some way because it was transboundary (i.e., the Maya Forest), and governments, which excitedly jumped at the possibility of tourist developments extending from the Mexican Caribbean toward Guatemala and Belize and other Mexican states (see also Laako et al., 2022). In his view, the problem is that the idea of the Maya Forest, originally destined for conservation, increased the governments' interest in exploiting the concept for other purposes – that of tourism. As a result, more Mexican federal states sought to join the project (for example, the *Selva Maya*-megaproject discussed in Chapter 3), but no longer for the purpose of nature conservation. Indeed, as also indicated in Laako et al. (2022), the interviewees on the Mexican side predominantly linked the Maya Forest concept to the Maya Train megaproject of the 2018–2024 Presidency. For them, the Maya Forest is basically part of the Maya Train project. Later, the conservationists interviewed by Laako in Petén and Belize mentioned government plans and talks about extending the train there. Most interviewees preferred to remain silent about these plans. A few of them expressed worry about the "danger" of this expansion for the Maya Forest owing to the destructive consequences of the Maya Train and not being able to "compete with Mexico for this massive tourism". Some also indicated understanding that such an opportunity could mean economic development and connectivity for the people and landscapes of the region.

During Laako's research on the Maya Forest between 2019 and 2023, conservationists tended to be reluctant to address questions of tourism, although the field observations clarified two parallel tendencies. On the one hand, conservationists tend to promote various forms of eco-tourism: bird watching and wildlife observation; visits of scientists, students, and selected adventurers to biological stations; and different kinds of tours of villages and/or communitarian tourism. On the other, there is the broader, national, and multinational tendency evidenced in major infrastructure

Image 2.5 A Maya Trail in a Protected Area
Source: Hanna Laako, 2022

projects, such as the Maya Train and the roadbuilding to the Caracol Archaeological Site in Belize, which have created both tensions and entanglements with conservation organizations. One of the entanglements has to do with the (re)production of the Maya, actively used by both, but with different intentions and aims.

When writing these lines, one of the most politicized topics in the context of the Maya Forest is the Maya Train (*Tren Maya*) mentioned above, which refers to the megaproject of the Mexican Presidency 2018–2024, inaugurated in 2024. This train extends from the tourist hotspot of Cancun to southern Quintana Roo to the remote forested and well-conserved but also waterless ruins of Calakmul in Campeche and Palenque in Chiapas, and connects to the long-ignored waterlands of Tabasco and the Yucatán. Currently, the topic is so politicized that it is difficult to obtain or communicate objective information. People are also scared to talk. It has not been possible to find exact proof that clearly shows whether the development of the Maya Forest concept was exploited or appropriated in the Maya Train megaproject and to what extent, although Laako's interviewees hinted in that direction. The book (Volume 1) by Martínez Romero et al. (2023) is one of the best on this topic, critically analyzing the impacts and origins of the train project from various angles and scientific disciplines. The book originates from a request made by the Mexican National Council of Humanities, Sciences and Technology (CONAHCYT) and the federal government's agency and was later censured by CONAHCYT and it was eventually brought out by an independent publishing house. Moreover, the documentary *Mayapolis* by Renaud Lariagon (2023), accessible on YouTube, which seems to be a result of French–Mexican scientific collaboration, explores tourism and the urban expansion on the Peninsula of Yucatán against the background of rural gentrification with links to the Maya Train, although it avoids assessing it directly.

In synthesis, these publications and Laako's research and interviews highlight three aspects related to the Maya Train. First, the conservationists' preoccupation with the loss of biodiversity in the Mexican Maya Forest as a result of the train. The threats are many, but in particular, there are worries about the subterranean waters and *cenotes* (sinkholes), given the karstic soil: will they become contaminated and destroyed by the train and what comes with the train? Moreover, attention is drawn to the situation of the Calakmul Biosphere Reserve, which is one of the best conserved and still holds some endangered jaguars. The Calakmul area has also provided connectivity with the best-preserved forest on the Guatemalan side. Some conservationists call this the "heart" of the Maya Forest. The area has been well preserved precisely because of the difficult access. At the same time, the lack of water resources, infrastructure and community involvement in the tourist expansion plans, as well as the militarization of the area (the Mexican military, for example, intends to build a hotel in Calakmul) have generated worries. While there has been an emphasis on the Calakmul area, many other locations along the Maya Train-route face the same situation. Hopes are high, even nostalgic, when related to the memory of past trains that coincided with economic growth and opportunities, but no infrastructure exists to receive the estimated numbers of tourists.

Second, doubts have been raised about the vague and opaque geopolitics behind the Maya Train owing to the government's plans for "territorial reordering" and the creation of new regions for tourist-, energy- and agroindustry by means of "12 locations" and "19 stations", which will "form part of poles of development", also defined as "sustainable communities"; this suggests urbanization and territorial transformation with unclear impacts in relation to organized crime, socio-environmental conflicts, migration, and land rights (Martínez Romero et al., 2023). Since the beginning of the megaproject in 2018, the cases of violation of Indigenous rights issued by the UN have been numerous (Martínez Romero et al., 2023, p. 19). Among others, Ceceña and Prieto (2023) have mapped the Maya Train route, indicating the locations of major biodiversity, minerals, oil, and access to the sea by the new train routes while blocking migration routes, which suggests that the Maya Train strategy is more than a railway with Maya-entitled railcars filled with happy tourists.

Finally, and as highlighted by the *Mayapolis* documentary, concerns have been raised about rural gentrification, a key issue affecting the Yucatán peninsula and entangled with tourism. The effects are varied. With tourism and wealthy affluent new inhabitants visiting or moving to different locations on the peninsula, such as Mérida or Tulum, and with increasing prices and costs, local residents cannot survive. Land speculation for the construction of hotels and residential areas has been increasing around the Yucatán Peninsula (Torres-Mazuera, 2022). At the same time, the political economy has changed from the traditional, socially owned *ejido*-system based on subsistence and livelihoods in the forest to the service-sector in tourism. Again, some sources have highlighted the ways in which the people in Yucatán view the train as a possibility for connectivity and economic opportunities.

Some issues remain to be explored as ongoing dynamics, in parallel with many other factors such as the government's *Sembrando Vida*-program, which seeks to enhance agroforestry in the region and the cultivation of fruit trees – aiming at a forest garden of sorts, at least on paper (Cano, 2024). These are new programs, the effects of which are yet to be examined. What is certain is that the Mexican Presidency 2018–2024 has taken its obligations seriously to bring "development" and make a difference to "forgotten, marginal and peripheral southern Mexico" (Martínez Romero et al., 2023).

Encounters with the Mayas in the Maya Forest Waterlands

> The Maya culture was born here. Mayas did not arrive here. People arrived here. Families arrived that later became the Mayas and these Mayas formed the culture. Formed this way of seeing, of feeling, of hearing, of relating and coexisting. (…) Each one finding their own way of life. Finding themselves in the water. Finding themselves in the earth. Finding themselves in the plants. Finding themselves in the rain. Finding themselves in medicinal plants. Finding themselves with the animals. Finding themselves with the birds. As these families began to find

> themselves in nature, belief was born, history was born, memory was born. (...) This way of the world that we now have as Maya.
>
> Pedro Uk, Buctzotz, YucatánSky Richards & Andres Kruger (2023), *El Tren y la Peninsula*, Documentary Film (21:26–21:30 min)

The new documentary film *El Tren y la Peninsula* cited above in the words of Pedro Uk has the merit of narrating the perspectives emerging from the actual contemporary Mayan territory subject to transformation in the context of the Maya Train (Richards & Kruger, 2023). What we learn about the Mayan perspectives on the Maya Forest continues to be little; however, in this unique film, we can hear and learn at least something.

Among other things, we are told that the train business is nothing new for the inhabitants of the Yucatán peninsula. The film starts with narrating the history of *henequen*, a plant used to produce textile and also called the "green gold", which was exploited in Yucatán for almost 200 years and which also brought about the previous railroad on the peninsula. While this period evokes nostalgic memories of the connectivity provided by the train, it is also remembered for the enslavement of the Indigenous people in the *henequen* business. However, as the *henequen* industry faded, so did the train, and the remains of this period have now been claimed by the jungle – the territory where the Mayas have lived for over 3,000 years.

The Mayas presented in the film have been told that the new Maya Train will have a positive impact on them. However, they wonder if the train cars will actually bring the hospitals, schools, serenity, and dignity they are looking for. Thus, the "development" promised by the megaproject is not clear. Instead, what these Mayan voices do know is how "development" has worked out in Cancun, where the Mayas now live as people's servants while big companies are seizing the land. As the cycle of *henequen* eventually came to an end, the same is likely to happen to the contemporary cycle of tourism. The trains have come and rusted away before; the fear, instead, is that the next cycle will impact the water. The Yucatán territory, according to the views expressed in the film, is considered the biggest source of underground waters and subterrain rivers in the world. For the Mayas in the film, the underground waters are also the places of the gods – the dark space – where the world was created. Will the veins of the *cenotes* be closed, and the water stop flowing? How will this affect what the Mayas have – the second largest jungle on the continent, the Maya Forest? (Richards & Kruger, 2023).

It is far from an easy task to address the final pillar of the (re)production of the Maya in the Maya Forest, namely, the Mayas themselves, as this is the Mayanism we hear less of. What kind of insights and encounters regarding the Maya Forest can we gain from the perspective of the Mayas themselves?

As indicated above, "Maya" is a scientific construct, which has later also been adopted by the self-defined Mayas. How have the Mayas called themselves? They have referred to themselves in many ways, as there have been Mayan groups at different times in different locations and circumstances. The

Mayas of the Yucatán peninsula, the *Mayab*, have been perhaps the most obvious people to be called Mayas. Many others have used names in their own languages referring to their particular community and group, occasionally calling themselves the "true people" or the "people of maize". The Mayas of the Mayan polity in the Chan Santa Cruz have called themselves *Cruzobs* (Castillo et al., 2006). Many others have called themselves by their linguistic group: Tsotsiles, Tseltales, Choles, Q'eqchi', and Mopan. They may have called themselves also according to their municipality or locality: Cobaneros and Lacandones. They have also called themselves Mexicans, Belizeans, and Guatemalans.

According to Wainwright (2008), we, as outsiders, should not even try to define the Mayas, but leave it to them to define themselves, and he may be right. The coloniality related to the contemporary Mayas and Indigenous people in general has to do with being defined by others, especially Westerners. Even the well-intentioned among us fall into coloniality when (re)producing essentialist, frozen, and often also romanticized perceptions of the Mayas, which reflect more our own needs and projections than those of the contemporary Mayas themselves. Over the decades, activist circles and anthropologists have been engaged in ferocious battles over *authenticity*: who counts as more authentic Mayas/Indigenous? Frequently, the essentialist positions have ended up eroding Indigenous struggles by dividing and excluding rather than supporting and helping. For example, urban Indigenous people have often been considered less Indigenous than rural ones. While Indigenous people may have their own dynamics related to rural-urban contexts and the fates of communities and migrants, outsiders often reject urban Indigenous people simply because, in their minds, the Indigenous only pertain to certain rural landscapes.

The same applies to describing the Mayas as forest people or *milpa* people; while these might be adequate and useful categories, they might equally result in unexpected problems. For example, Wainwright (2008) notes that the Mayas in southern Belize had been written in Belizean history as scattered and unsettled and thus a resource for development. In his view, the colonial perception of the Mayas as forest people was linked to racist connotations about "aborigines who cultivate maize somewhere in the depths of the forest and fatten pigs" (Wainwright, 2008). While the contemporary discussions about Mayan agroforestry and forest people may have completely different undertones and connotations, is it possible that we fall into the same traps today?

Our approach to understanding the Mayas used by the Mayas themselves has more to do with a political-collective identity and strategy to defend Mayan territories and rights. According to Scott (1990), resistance to domination and existing power relations emerges first within the same hierarchical frame in a seemingly acceptable social form, although from the beginning, this seemingly adaptable discourse may have already gained a completely different meaning behind the scenes. In other words, *hidden transcripts* – meanings, dissimulation, and management of appearances in the public discourse prior to any open rebellion – are always at play. As we also know from the theories of social movements (e.g., Tarrow, 1998), windows of opportunity opening in the power structures may also catalyze cycles of protest.

While the Mayas have long rebelled, resisted, and defended their territories in multiple forms, a particular window and strategy opened in the 1990s: that of Indigenous people' rights. In the case of the Maya Forest space, two particularly strong manifestations emerged – the Guatemalan Mayan movement and the Chiapanecan Zapatista movement. Both have long histories in land-right struggles and left-wing insurrections, which had predominantly built a political identity via the category of peasants. Yet, the 1990s Indigenous emergence (Brysk, 2000, Bengoa, 2000; see also Engle, 2010) shook the earlier categories and emphasized Indigenous rights. The Indigenous emergence is often considered to have been catalyzed by the commemoration of 500 Years of Resistance in 1992, although it is rooted in earlier developments in Latin America and the Americas in general, which gave rise to the collective identification as Indigenous people and new mobilizations as Indigenous movements.

In both cases in Chiapas and Guatemala, ruptures divided or shaped the political strategy toward the framing of Indigenous people's rights. In the case of the Zapatists, it was the encounter with civil society posterior to the uprising in 1994 that led, among others, to a rewriting of the colonial connotations of the "Indian/Indigenous" in Mexico, the arduous negotiations with the government to achieve the San Andrés Agreements, and the formation of the autonomous Zapatista territory. In the case of Guatemala, the rupture with the revolutionary wing and the adoption of the "Maya" in the 1990s transformed the Mayan movement into a political actor contributing to the framing of Indigenous rights. According to Bastos and Camus (2004), the Guatemalan Mayan movement opted for the term "Maya"; this is because they considered the notions of "Indian" and "Indigenous" as colonial impositions and thus characterized by subordination, whereas the term "Maya" was collectively chosen because it was rooted in a distinct political context that suggested historical continuity. Despite deriving from Archaeology, Anthropology, and Linguistics, the adoption of "Maya" allowed to take critical distance from being defined as peasants (Bastos & Camus, 2004).

The 1990s Indigenous emergence coincides with another emergence: that of the concept of biodiversity and the Maya Forest. We (Laako & Kauffer, 2021, 2022; Laako et al., 2022a, 2022b) have previously narrated how these two initially clashed over the slash-and-burn methods. The topics most extensively explored by scholars are the violent dislocations of peasants in protected areas in Guatemala (e.g., Ybarra, 2018) and green land-grabbing. The conflict between biodiversity conservation and the Mayas in the Maya Forest shaped these relations and, in the best cases, also shifted the conservationists' approaches. In Laako's research, several conservationists have alluded to the context of the 1990s as transformative, when the earlier perceptions changed toward collaboration and the inclusion of communities in landscape connectivity.

This issue has another layer: the demand to decolonize scientific research and the researchers' relationship with the Indigenous communities. This demand has been particularly marked within the Zapatismo that sparked a strong anti-academic tendency in Chiapas in the 2000s. In many contemporary Indigenous

Image 2.6 The Zapatistas' Silent Protest on the Supposed Mayan Apocalypse Day 12/12/12, Chiapas, Mexico
Source: Hanna Laako, 2012

movements and communities in the Maya Forest, academic research is still treated with justified suspicion.

These different layers of politicization related to the Mayas, science, and biodiversity may have culminated in the conflict between the Maya International Cooperative Biodiversity Group (Maya-ICBG) and the Chiapas Council of Traditional Indigenous Doctors and Midwives (COMPITCH) at the end of the 1990s. The case politicized bioprospecting and brought it to global attention. Bioprospecting refers to the search of products originating from different kinds of biological resources, such as plants, animals, and microorganisms. These are then intended to be commercialized for the benefits of society. According to Takeshita (2000), bioprospecting involves multiple stakeholders, from drug companies and scientists to ethnobotanists, governments, and organizations. Bioprospecting often involves Indigenous people, especially in biodiversity-rich regions such as tropical rainforests. It also tends to impact on Indigenous people as they are often enlisted by bioprospectors as local collaborators owing to their knowledge of their lands: they are considered facilitators in the discovery of valuable natural resources and regarded as information providers and executors of biodiversity conservation (Takeshita 2000). The 1992 Convention on Biological Diversity sought to address the problems deriving from bioprospecting by endorsing the rights of Indigenous people to receive an equitable benefit for sharing their knowledge. In practice, these endorsements have not been easy to implement, and many Indigenous people protest against what they consider to be appropriation.

The case of ICBG-COMPITCH was, however, even more complex. The ICBG was composed of a consortium involving the University of Georgia, El Colegio de la Frontera Sur (a public research institute in Chiapas), and MolecularNature Limited, a private British pharmaceutical company. They were seeking natural products with medicinal properties and, for those purposes, sought to collaborate with the Mayan communities in Chiapas (Nigh, 2002). The problem occurred with prior informed consent, which suggested landowner representation, although the products' commercialization extended to Indigenous knowledge embodied in the potential products derived from knowledge structures and livelihood practices through which Indigenous people have managed and preserved the biodiversity in their territories (Nigh, 2002, p. 461). Thus, according to Nigh (2002), an individual owner of a given piece of land in which the collection is made is not the sole owner of the knowledge related to those species, and therefore could not give informed consent as a sole representative of the wider community. ICBG sought to arrive at this sort of broader consent with COMPITCH; however, negotiations failed owing to the lack of a Mexican regulatory framework for bioprospecting. As there was no such framework in place, ICBG eventually decided to pursue its project based on the existing framework – that of direct consent from communities or parcel-owners. In other words, they used the same form as traditionally used by other researchers when approaching Indigenous communities for research, but in this case, it was a bioprospecting project. Meanwhile, COMPITCH became wary of the regulatory vacuum and solicited international advice from the Canadian Rural Advancement Foundation International (RAFI), which found that the ICBG project was potentially damaging the Mayas' interests (Nigh, 2002). An open and global controversy ensued when the ICBG project was reported as "biopiracy" and the communities advised not to sign informed consent. No agreement was reached as no regulatory frame was provided by the Mexican government, and ICBG did not accept COMPITCH's request for exclusive intellectual property, which was, in the view of COMPITCH, based on the principles of common heritage and public domain.

Such controversies as the case of ICBG-COMPITCH in the interstices of Indigenous rights, heritage, biodiversity, commercialization, and science are also key challenges within the Maya Forest landscape, a central part of its being *Maya* and *Forest*.

The ICBG-COMPITCH case highlighted important issues of intellectual property, along with many others that involve biodiversity and the natural resources of the Maya Forest. The Global Atlas of Environmental Justice , for example, lists cases of Mayas defending their lands against forest concessions (discussed below) and against oil exploration in the Temash-Sarstoon National Park in Belize. In Petén, the Atlas identifies cases of displacement of peasants from the Laguna del Tigre and Sierra del Lacandón National Parks (see also, e. g., Ybarra, 2018) as well as successful community mobilizations against touristic megaprojects in the Maya Biosphere Reserve. For southern Mexico, the Atlas shows many cases, such as the protests against hydroelectric dams in the Santo Domingo River (see also Laako & Kauffer, 2021; Chapter 4) and the Usumacinta River, against REDD+ pilot in Chiapas and against the Maya Train in Calakmul, the Mayan beekeepers against Monsanto (see also Peréz Ruíz, 2018), and several cases of Mayan communities against eolic parks and windmills. In Quintana Roo, conflicts have arisen about tourist projects threatening livelihoods and biodiversity.

However, these environmental conflicts involving Mayan and Indigenous communities do not automatically suggest a conflict between conservation/science and Indigenous rights. As shown before (Laako & Kauffer, 2021, 2022), in many cases, communities and conservationists have formed coalitions against extractivist megaprojects. In fact, it is not always possible to distinguish between such categories as "conservationists" and "Indigenous communities", as many conservation organizations include Indigenous people, and Indigenous communities are also conservationist. Our research suggests that it would be a mistake to automatically assume a conflict between conservation and Indigenous people in the Maya Forest because they interact with each other. However, our suggestion does not mean that these links and relations should not be critically examined, as shown in this chapter.

In the following "insights", we address the many subtle ways in which the *Maya* of the Maya Forest waterlands appear today as part of the region's territorial transformations. The first insight, emerging from Kauffer's fieldwork, sheds light on the history of the Usumacinta River and contemporary tendencies as part of the Mayan narratives. In the second insight, emerging from Laako's fieldwork, the contemporary struggles and entanglements between protected areas, forests, and the *Maya* are illustrated in the case of the Columbia River Forest Reserve.

Insight 1: Mayan Rivers Then and Now: Waterlands from Mayanism to Tourism

By Edith Kauffer

According to the archaeological and historic literature, one of the most traditional "Mayan rivers" of the Maya Forest waterlands is the Usumacinta River. Identified as "the Mayan river *par excellence*", the Usumacinta River is also referred to as "the ultimate wet boundary" and "a cascade of languages, individuals and cultural expressions flowing from a source of clear Mayan affiliation" (Ruz, 2010, p. 7). This "main natural track" to the lowlands of Tabasco, Chiapas, and Northern Guatemala (Obregón Rodríguez & Liendo Stuardo, 2016, p. 21) is considered the gateway to the cradle of the Mayan civilization and an ancient route of communication between the Gulf of Mexico and the Caribbean Sea.

The Usumacinta River – described in more detail and from diverse perspectives in Chapters 3 and 4 – has been a major watercourse for archaeologists interested in the numerous sites but also for explorers (Obregón Rodríguez & Liendo Stuardo, 2016) from the nineteenth century onward. Downstream in Tabasco are the lowlands, with either 40 or 200 archaeological sites (Ruz, 2010), depending on the source. According to Scherer and Golden (2012), upstream in Chiapas and Guatemala, there were 23 sites, and about 31 places (Golden & Scherer, 2013) in the surrounding area of Piedras Negras and Yaxchilán, two kingdoms competing and flourishing between the sixth and ninth century AD. More recently, Scherer et al. (2022) presented an updated map of the Late Classic Period in proximity to Piedras Negras with two royal sites, five secondary centers and about 40 other sites, based on a lidar survey.

For this reason, the Usumacinta River could be considered a pillar of the Mayan narrative, from pre-Hispanic times up to the most recent government programs focused on tourism and orientated toward this "Mayan river". In my perspective, the river that structures a whole region according to Ruz (2010) dominates the waterlands that we will analyze in the next chapters, primarily consisting of a communication channel between the three transboundary corridors defined in our study: the Lacandon Forest Waterlands in the south, the Laguna del Tigre-Gulf Riverlands in the north and the Caribbean Karst Waterlands in the eastern part of the region. This communication axis depends on the existence of the Usumacinta River and other waterways, and on land connections that once again allude to the close links between waters and lands, not only with regard to nature but also to local societies.

The reality of the Usumacinta River as the major and unique communication route in the past and present may be considered overstated owing to the fact that parts of the main watercourse have been non-navigable (see Chapter 3), particularly from the Piedras Negras archaeological site, currently located on the Guatemalan side of the international river, up to Boca del Cerro in Tabasco downstream in the north, as well as at some places in the south, such as the Anaité or Chicozapote rapids. The difficulties associated with the Usumacinta's rapids, which have been described from pre-Hispanic times up to today, create a need for a combination of sidewalks or land routes around the river to access the region. During his early twentieth-century travels, Maler (1901–03) mentioned this situation, and more than a century later, Scherer and Golden (2012) also described the impossibility of traveling by canoe along the river during pre-Hispanic times, and explained that traveling through the rapids remained dangerous, even at the beginning of the twenty-first century. My own team had an accident in January 2018 at the San José Grande Rapids, in which nobody was injured, but which resulted in damage to property.

> When Maler explored the region, however, it was largely forested and only sparsely occupied by logging encampments and scattered Lacandon Maya communities. Travel for Maler was slow and arduous overland, whether the paths of the wood cutters or bush-whacking through the forest. Travel

> by canoe was a little better on the fast-flowing Usumacinta River. Although, it is often assumed that the majority of travel, and particularly, the movement of bulk goods through this region in pre-modern times was by canoe on the Usumacinta, we find this doubtful.
>
> Scherer & Golden, 2012, p. 13

Archaeologists thus point to the role of rivers, such as the Usumacinta, as routes in a necessary combination with terrestrial ways of access. Obregón Rodríguez and Liendo Stuardo (2016) explained this important aspect in detail describing the physical characteristics of the river and the topography of the whole region denominated "High Usumacinta" from an especially Mexican perspective – one that tends to forget the upstream part in Guatemala – in opposition to the "Low Usumacinta", which covers the downstream region from Boca del Cerro to the Gulf of Mexico in the Mexican states of Tabasco and Campeche (see Chapter 3). The team of Golden and Scherer, which was still working in the region in 2023, called the area the "Middle Usumacinta River Valley" and concluded that, in the past, it had been impossible to do the whole journey on the river. Thus, the idea of the "naturally vertebrate by the same river and culturally marked by its reiterated Mayan affiliation is still observable. Major River, Mayan River" expressed by Ruz (2016, p. 19) must be correctly specified as a mixture of river and land routes in the Middle Valley. Canter (2007) previously published a detailed analysis of the Usumacinta River with a map, in which he proposes a combination of river sections and portage routes. During their archaeological travels, Scherer and Golden (2012) followed Maler's route, a journey that took place in 1901, and reached exactly the same conclusion.

> As we advanced, the river became more sinuous and faster, the mountains steeper and the pass narrower. Soon we were squeezed between high grayish rocks whose edges were vertically entering the water; some of them looked like crenellated towers or half-collapsed walls. Before we knew it, the cliffs were getting closer, the space was narrowing even more, the sun was disappearing and the shadow of the *sierras* ranges was spreading over us like a veil.
>
> Morelet, 1857/2015

Trying to travel upstream from Tabasco to Chiapas, Morelet (1857/2015) described his first encounter with the rapids. All these historical testimonies and recent research evidence that it was never possible in the past to complete the journey on the river, and although it is possible now to do so with engine boats, this journey can be especially dangerous because of the San José El Grande Rapids, which are the most difficult stretch to pass through. Additionally, Canter (2007) and Scherer and Golden (2012) explain that without an engine, it is much more difficult to navigate upstream than downstream. Nevertheless, navigation on parts of the Usumacinta River was a reality, given that attachment marks and rope grooves reveal the existence of river docks and ports (Canter, 2007; Canter & Pentecost, 2007) in different locations such as Piedras Negras and Yaxchilán, with the former now located on the Guatemalan riverside and the latter on the Mexican side.

Archaeologists assume that there were conflicts and constant military confrontation to control trade routes between the kingdoms of Piedras Negras and Yaxchilán, situated on opposite riversides (Obregón Rodríguez & Liendo Stuardo, 2016; Scherer & Golden, 2012) until their collapse, which has been interpreted as the result of their growth and as a consequence of complex political and economic processes (Golden & Scherer, 2013). The location of the two capitals ensured strategic control over the river route as well as the connections with land routes and with sites of lesser importance, which were also necessary to the establishment of both river and land communications. The capitals were hence protected by rapids and connected to land portage routes. For this reason, Obregón Rodríguez and Liendo Stuardo (2016) assert that the river was fundamental to the pre-Hispanic occupation of the region. There was an ongoing struggle for control of the river and for the domination of the region's smaller population centers between both kingdoms, which tried to incorporate peripheral sites (Schroder et al., 2017) to strengthen their own defense but also to secure portage routes and safe crossings of the river. My recent fieldwork on sedimentary areas in the Usumacinta River revealed that the gravel and sandy beaches coincide with the landing places for barges, identified as ancient and current ports by archaeologists (Schroder et al., 2017). Consequently, the dynamics of waterlands that refer to the fluidity of these "human-water-land systems" (Camargo, 2022) reveal indissociability and entanglements. In this case, the interconnections, which depend both on the river levels and thus the seasons, and on the location, whether upstream or downstream, are also a result of the political and military objectives of the ancient riparian kingdoms that have contributed to their creation, transcending the natural boundaries between water and land. Water and land routes have been closely integrated to ensure their domination of the Mayan waterlands.

Despite the difficulties of transport in the Middle Usumacinta River Valley, the river has been used by explorers and by companies extracting timber and chewing gum (Obregón Rodríguez & Liendo Stuardo, 2016), as presented in Chapter 5, notwithstanding the existence of ten rapids including those located on the international river, of which six are dangerous, according to a highly detailed map of the watercourse (Canter, 2007). At the end of the nineteenth century, in the context of the border dispute between Mexico and Guatemala, the Usumacinta River was also visited by people commissioned by both governments to draw up maps and carry out site surveys (Obregón Rodríguez & Liendo Stuardo, 2016). Consequently, the mix of river sections and portage routes has always formed the spine of the region, except for timber exportation, which used the whole watercourse from the upper tributaries in Guatemala to the Mexican port of Frontera, located in the Gulf of Mexico.

In contrast to the "Middle Usumacinta Valley", the lower Usumacinta has been and remains totally navigable. During Spanish colonial times, downstream Usumacinta was known as "the region of the rivers" (*Región de Los Ríos* in Spanish; Ruz, 2010), characterized by thick forests, swamps and wetlands, and numerous rivers. The colonial name of "the region of the rivers" corresponds to the lower Usumacinta, which extends from Boca del Cerro to the Delta, including its downstream tributaries, the Palizada and San Pedro-San Pablo Rivers, which flow directly into the Gulf of Mexico. Today, the Palizada River is no longer navigable, but it was during the nineteenth century, as mentioned by Morelet (1857/2015).

After the colonial era, the region of rivers was converted into an administrative district of the state of Tabasco. Between the sixteenth and the beginning of the twentieth century, this region included all the riparian settlements and municipalities inside the region of the rivers or Usumacinta, although it has sometimes been ignored by the new administrative powers owing to its remoteness from the regional political center. According to Aguilar Palafox (n.d.), in 1947, del Águila and Bernado defined four administrative regions in Tabasco with the region of the rivers being one of them. In 1994, a regional reorganization of the state of Tabasco created two regions, the upstream Usumacinta of Tabasco, called the region of the rivers, and the downstream part of the river designated the region of the swamps, including the deltas of the Usumacinta and Grijalva rivers that merge about 25 km before emptying into the Gulf of Mexico (Secretaría de Gobierno, 1994). This administrative delineation is hence the place where the Maya Train begins its route and was inaugurated in 2024 by the Mexican President.

The Maya Train is based on the idea of the Maya Route, developed in 1988, and framed by the "Maya World", an organization created by the tourism authorities of five countries in 1992 and set up in 2005 in Tabasco to promote regional tourism. Continuing these previous initiatives, the Regional Program of Tourism Development of the Maya Rivers Route of the state of Tabasco, established in 2019, promotes a tourist program called "Mayan Rivers", created by the federal government and focusing on six projects.

One of them is the Usumacinta eco-cruise, which entails navigating the lower Usumacinta River along the whole stretch of the river in Tabasco, with activities for tourists linked to the river and nature. Other projects involve the construction of harbors such as the River Terminal "Boca del Cerro", associated with the Maya Train station of Tenosique and the protected area of the Usumacinta Canyon, the quay of Tenosique, the Unity of Tourist and Nautic Services in Balancán, the River Terminal, and a recreational and business complex in Emiliano Zapata, situated downstream as part of the planned infrastructure. There are also plans for an adventure tourism center and eco-hotel, "Río San Pedro" on the San Pedro River, a transboundary tributary of the Usumacinta River, originating in Guatemala and merging with the Usumacinta in Balancán. Downstream are a Manatee Interpretation Center in Jonuta, where local residents present manatees to tourists, and a quay in Frontera, at the confluence of the Usumacinta and Grijalva Rivers, which then flow together into the Gulf of Mexico.

The narrative of this program, announced in 2019 and always presented under the name of "Mayan Rivers", is based on the historical navigation of the river by the Mayas. The river is thus presented as an entry to the Maya World, and linked to nature and sustainability in reference to the presence of fauna and flora, even though there is no forest due to the deforestation of the area. Once again, Ecology and Mayanist narratives have been combined. The inclusion of the Wildlife Reserve Canyon of Usumacinta sustains part of the program in the most deforested area of the Usumacinta River. Six archaeological sites have been opened to the public, ten are susceptible to be exploited, and six riverbank municipalities involved.

The program was presented in 2019, and with the exception of the new protected area of Wahna', which was an older project promoted by researchers, none of the planned infrastructure has been built by June 2024. August 2023 saw the creation of the Wahna' Biosphere Reserve on the San Pedro Mártir River, a tributary of the Usumacinta River. It was awarded reserve status owing to the existence of the red mangrove (*Rhizophora mangle)* and some bird species. Wahna' is a Mayan word that means "river of the quails."

From ancient times to the most recent touristic projects, archaeological research in the Maya Forest waterlands refers both to an ancient civilization and to ecological and conservation concerns, even though the lowlands are deforested to a large extent. In this region, the links between fluvial and land routes present a major connection between water and lands. Navigation on the lower Usumacinta is a major issue and there has also been a recent project, initiated in March 2023, to dredge the Palizada River, which was initially not included in the Maya Rivers Program, but would be of strategic importance to improve river navigation and give access to tourism.

Insight 2: Co-Conservation in Forest Waterlands and the Mayas of Belize

By Hanna Laako

> It is difficult to carry out conservation of landscapes with the Maya people in it.It would entail the recognition of Maya land rights.
>
> Mayan leader, Belize, June 2023

According to Anaya (1998), the government of Belize, through its Ministry of Natural Resources, granted at least 17 concessions for logging on lands totaling around 480,000 acres in the Toledo District in the 1990s, which had a negative impact on the Mayan people. In 1996, the Mayan organizations took legal action challenging the logging concessions as a violation of Mayan rights. With logging linked to environmental degradation and conservation, from the Mayas' standpoint, the concessions had a fundamental impact on the ownership and control over these lands and resources, based on historical occupancy and ongoing customary land tenure.

Wainwright (2008) has extensively narrated the history of the battle over the Columbia River, which was the focus of this particular borderlands struggle and encounter involving complex colonial history, Indigenous territory with its forest people, modern nation-states and extraction, and forest waterlands. In Wainwright's (2008) account, the story is personified by soil biologist Charles Wright, in service of the empire in colonial Belize, who purchased land and established a farm in southern Belize. He also became the author of the book *Land in British Honduras* in the 1950s, which set in motion a particular developmental discourse on the Mayan people. According to Wainwright (2008), Charles Wright was also a Mayanist with a keen interest in the *Maya* as well as environmental issues, and who, despite being a colonialist himself, was also considered a defender of the Mayan cause and forests.

Without going too deep into the history and writings of Wright, which can be consulted and read in more depth in Wainwright's work, we only mention the main point here, that it was Charles Wright who "discovered" the Columbia River Forest as an important socio-natural space in his soil survey of 1953, which led to the idea of a forest reserve. In 1958, the colonial government, encouraged by Charles Wright, did indeed set aside 132,750 acres of wet tropical forest to establish the Columbia River Forest Reserve, where hardly any logging was done over the following 35 years, except for a modest amount of timber collection by the local Mayas (Wainwright, 2008). The Columbia River Forest Reserve comprises nearly 60,000 hectares and is home to six watersheds, the largest one being that of the Río Grande (Meerman, 2004).

In the 1990s, logging pressures increased in Belize, and concessions were given to various companies in the area of the forest reserve. Eventually, these developments coincided with another series of events – the emergence of the new Mayan movement led by Julian Cho and a new organization of the *alcaldes*. The logging concessions in the Columbia River Forest Reserve sparked off a broader demand for land tenure, based on the argument that the future of the Mayas and the forests were closely intertwined (Wainwright, 2008). National debates ensued on Belize's dependency on forestry revenues and questions about the Mayan land use as "a threat to the forest". Eventually, this encouraged the Toledo Maya Cultural Council and the Alcaldes Association to initiate their own mapping of ancestral lands and a lawsuit against the government. These endeavors were supported by Wright and several other scientists.

In addition to the lawsuit, the Mayan mapping resulted in the *Maya Atlas* (1997), created by 42 Q'eqchi' and Mopan Mayan communities in southern Belize and supported by the work of anthropologists, archaeologists, and geographers (Anaya, 1998). This map, a product of the collective Mayan effort and used in court, can be viewed on the website of Nawimaps (https://nawimaps.com/maps/maya-atlas/).

The mapping served to underpin the claim for historical occupancy and ongoing land tenure. According to Anaya (1998), the detailed account of the historical and contemporary land and resource use patterns of the Toledo Mayas went largely unrefuted by the government, although some arguments were presented referring to the Spanish conquest of the territory as well as an attempt to portray the contemporary Mayas as unrelated to the earlier Mayas inhabiting the territory. Yet, these were refuted by the evidence provided by the Mayas, which corroborated both the exclusive ownership rights over certain territories and the non-exclusive right to engage with subsistence and cultural activities farther away from the villages, based on customary patterns of migration with shifting cultivation, hunting, and gathering activities, which also involve sacred places in permanently forested areas. Despite the migration patterns and colonial displacements, the Mayan homelands were never abandoned, and they were later identified as southern Belize.

The map provided by Nawimaps illustrates an area that covers the contemporary Districts of Toledo and Stan Creek and includes the Columbia River Forest Reserve. The map also shows, for example, the Bladen Nature Reserve as the outer Mayan hunting grounds. Most of the coastal area from the city of Punta Gorda (PG) northward is not marked as Mayan lands, whereas the southern borderlands including the Sarstoon-Temash National Park are identified as "outer Mayan hunting grounds". The Sarstoon-Temash National Park was established in 1994 prior to consultation with local residents and was later designated as a Ramsar wetland site in 2005. Currently, the area is co-managed by the Forest Department of Belize and the non-profit NGO Sarstoon Temash Institute for Indigenous Management (SATIIM), created in 1997 by the Garifuna and Mayan communities in southern Belize.

These Mayan maps, as explained above, coincide with several contemporary protected areas, which result from the work of and are managed by many conservation organizations. One of them is the Ya'axché Conservation Trust, founded in 1998 as a consortium of local leaders with the aim to conserve the corridor connecting the forests of the Maya Mountains with the lowland forests of the Caribbean coastal plains, known as the Golden Stream Corridor Preserve. The Ya'axché Conservation Trust also co-manages the Bladen Nature Reserve and the Maya Mountain North Forest Reserve. Altogether, these lands total 770,000 acres and are called the Maya Golden Landscape, of which the part approximately extending eastward from the Southern Highway is identified as Mayan lands.

In my interview with the Friends for Conservation and Development (FCD), which co-manages the Chiquibul National Park along the contested border/adjacency zone (i.e., borderlands), the leader commented that the Belizean Forest Department, having only limited funds, had encouraged the FCD to expand their patrols toward the Columbia River Forest Reserve, viewed by the FCD as the broader Maya Mountain landscape (interview in San Ignacio in June 2023). In the leader's opinion, at the moment, only the FCD has updated data on what is happening in the forest reserve near the borderlands and next to Chiquibul, given that it conducts flights over the area several times a year. During these flights, the FCD has observed increasing logging and cattle-ranching from the Guatemalan side; it suspects the involvement of a new drug-cartel. Additionally, fresh water is more and more scarce on the Guatemalan side, which also augments the pressures on the Belizean side, holding more abundant water resources. As a result, the leader has suggested a system of co-management with the Mayan leader alliance in Toledo. This proposal of collaboration is based on the fact that these areas are designated as Mayan lands, but also have ties to the Guatemalan communities, which is why a joint effort could contribute to addressing these challenges affecting the forests and people – in other words, the Maya Forest – in the Belizean/Guatemalan borderlands.

The leader of the Ya'axché Conservation Trust shares a similar approach (interview in Punta Gorda [PG] in June 2023). She discusses the "expanding problem" of illegal logging, cattle-ranching, and poaching from the Chiquibul National Park toward the Bladen Nature Reserve, which borders both the Chiquibul National Park and the Columbia River Forest Reserve in the west. While the Ya'axché Conservation Trust relies on patrolling forces, its leader says that co-managing the Forest Department is complicated as the rules and responsibilities fall on them to ensure conservation while receiving no funding from the government. Thus, the Forest Department has no presence in the reserve and the Ya'axché Conservation Trust – and the FCD – deal with the challenges as best as they can while engaging in fundraising for those efforts.

In her view, there is a similar situation in the neighboring Columbia River Forest Reserve. The leader is acutely aware of the 2015 Court Intermediary Consent Order regarding the Maya customary land use, which also applies to the Bladen Nature Reserve in addition to the Columbia River Forest Reserve. She says that no agreement has been reached and they have not been given a model on where and how the protected areas fit into the framework of the Mayan land rights, while problematic and hostile tensions within the protected area are affecting both the patrols and the communities. Work on defining the boundaries, but also the denial of the protected area are ongoing, and while the government is working on the legislation, it has not consulted the communities as it should have done. In her view, the Mayan communities depend on the forest and have always been its custodians. Yet, both sides face challenges as to how to adopt to and negotiate their different needs. Additionally, both FCD and Ya'axché are part of a broader inter-institutional meeting set up to deal with the border situation, particularly the escalating drug-cartel issue during the past three years.

In the Mayan leader's view, it must be made clear that these remaining forests are situated within the Indigenous people's lands, which have been fought over for the past 20 years and require a fine balance between environmental protection and respect for the Mayan rights (interview in PG in June 2023). He favors the model of human rights to conservation and climate change, including the forest management by the regional, Latin American, and Caribbean Indigenous people. This human-rights approach is fundamentally aimed at supporting the Mayan communities and rights and seeks to explore the frameworks and tools together with them as part of the solution. The Mayan leader considers the predominant model of conservation a "nightmare" in the sense that it usually involves the idea that Indigenous people need to be taught to take care of the environment, whereas the most intact forests are found precisely in Indigenous territories.

However, according to him, the principal problem of this predominant model of conservation is that it is based on the assumption that the lands and territories are in the hands of the government, although the integrity of the natural resources is never guaranteed. As this type of conservation requires fundraising, a kind of tokenism appears based on "Maya this – Maya that", while there is no real engagement with the Mayas of the Maya Forest. In this sense, the Maya Forest is a convenient platform to attract funding, although he adds that he personally has nothing against any of the Belizean conservationists. Yet, to really include the Mayan people in landscape conservation would require the recognition of Mayan land rights, which, again, would mean that protected areas do not revoke customary rights.

In this sense, he considers co-management a failed model as, in reality, it is a government-owned model. While the co-management model may allow conversation and engagement with people, it does not change the land-ownership scheme in any way. From the perspective of the Mayan people, the problem is this: they feel that the Columbia River Forest Reserve belongs to them and not to the government. If they, as Mayas, were to accept a co-management plan, it would inevitably suggest that the area belongs to the government and not to the Mayas. Thus, prior to any co-management plan for a protected area, it should be acknowledged that the territory in question belongs to the Mayas – as confirmed by the ruling of the Caribbean Court of Justice.

The Mayan leader takes, as an example, a recent case from Panama, where the Indigenous Naso people won land rights and the role as guardians of the environment along the Teribe River (e.g., Dorman & García, 2021). In this way, the land was first recognized as Indigenous land and the Indigenous as guardians for its conservation, not responsible to the government but to the communities. In the Mayan leader's view, people need to be trusted. The Mayas consider that they, as people, are healthy when their forests are healthy.

Second, the Mayan leader criticizes the government's co-management model as risky because government interests shift and change, and resources may thus be exploited despite the borderings and fences – as it was the case in the Columbia River Forest Reserve in the first place. He argues that space for biodiversity is often shrinking because of government policies, when, for example, the policy is to first build a road and then to invest in jaguar conservation as the animal is threatened by the road.

Nevertheless, the leader comments that they, as Mayas, are aware that the spaces of Indigenous people are always contested. These spaces are now often imposed by jurisdictions, which, however, do not eliminate any existing connections and the Mayan ways of being. Thus, the Mayas of Belize do not seek to get rid of national boundaries, but to strengthen connections in order to survive – they have always been interested in transboundary relationships.

In this sense, he acknowledges, for example, the role of the FCD in handling the pressures in the transboundary space occupied by both communities and people with a shared history of displacement and lack of access to land, as well as the diminished health of the forest, now also challenged by the explosion of organized crime. Yet, relationships and engagements are enabled across the border. The leader emphasizes that he values the collaboration and efforts undertaken by the FCD and Balam in Guatemala to form a strategic alliance in order to alleviate the pressures on land and resources.

The case of the Columbia River Forest Reserve, at the crossroads of the mappings of contemporary Mayan rights, protected areas and conservation and government policies in borderlands bring to the fore at least three parallel issues. One is the fact that the seemingly uninhabited Columbia River Forest area is a hotspot of contemporary Mayan struggles, lands, and encounters with national and (post)colonial powers, which has even given rise to the contemporary Belizean Mayan movement. This challenges the notion that there could be no borderlands in "uninhabited" areas (Readman et al., 2014). Second, the case of the Columbia River Forest Reserve shows a strong entanglement of bio- and geopolitics in the Mayan borderlands, with dynamics resulting from unresolved border disputes and transboundary pressures, affecting protected areas managed by conservationists. Finally, the case of the Columbia River Forest Reserve suggests the need to critically examine the co-management model, which has often been offered as a solution – even one addressing decolonization – to conflicts about conservation management. As pointed out by the Mayan leader in Belize, while the co-management model may help to include more people and foster mutual learning and collaboration, it does not change land ownership, which, in the view of many Indigenous people, lies at the heart of decolonization (e.g., Tuck & Young, 2012).

Conclusions: On the Things Left Underfoot

The Maya Forest has been subject to active regionalizations and borderings in its contemporary history. A substantial part of those regionalizations and borderings corresponds to conservation work, which eventually gave rise to the concept of the Maya Forest in the 1990s.

However, the Maya Forest is more than the sum of its ecological borderings and protected areas. It is a particular space – discursive and physical in the form of protected areas – created in the interstices of Archaeology and Ecology working toward new approaches involving the Mayas as forest people and the conservation of the ruins and species in the remaining rainforest. In other words, the Maya Forest is a place of encounters and a place of work for scientists, organizations, and often also for communities. In a similar vein, it needs to be emphasized here, as in Chapter 1, that in their coming-together to understand the Mayas as forest people, and to safeguard the ruins and surrounding forests, both conservationists and archaeologists have also actively applied the concept of biocultural diversity, which attempts to respect the needs of local inhabitants.

In addition, the Maya Forest as an eco-borderland allows shedding light on what is often excluded or silenced, i.e., left underfoot of the predominant narratives, which represents the primordial focus of Borderlands Studies. In the case of the Maya Forest, this includes the role of archaeologists and conservationists establishing a transboundary working space consisting of protected areas such as heritage sites and biosphere reserves forming the so-called "Maya Arch", which encompasses both ancient Mayan ruins and surrounding forests. As such, the Maya Forest might then have more to do with scientific collaboration than specifically with the Mayas, such as the Belizean Mayas in the Columbia River Forest Reserve actually do with their forests.

The Maya Forest is a result of, and part of, the histories of Mayanism, generally defined as the study of the Mayas. Yet, as such, it is linked to the broader (re)production of the concept of the Maya that builds on and appropriates Mayanism. For these reasons, in this chapter, we identified three pillars that sustain and reproduce the *Maya* in the contemporary Maya Forest: scientific Mayanism, the tourist industry of mayanization, and the Mayas themselves as part of the strategies emerging from the Indigenous movements. There are groups who do self-identify as Mayas to defend their Indigenous rights and territories. The latter are equally part of the Maya Forest landscape, often excluded from the predominant Maya Forest narratives. Thus, in this chapter, we have suggested that such "uncomfortable" issues as bioprospecting, land rights, and collective knowledge must pertain to the landscape and territory that we like to call "Maya".

These are also transformative processes in the contemporary Maya Forest waterlands, which shape and change the territories and their borders. Along these forest waterlands, the *Mayan rivers* that flow from the highlands to the Mexican Gulf and Caribbean Sea are part of a long history of human settlement, colonization, and livelihoods, as well as a way of transport, exchange, and connections, especially in the floodplains located in the three countries. Today, the rivers and waters are also part of a large tourism business that has developed in many formats – from eco-tourism and homestays in Indigenous communities to the seaside resorts and forest eco-lodges distributed throughout the region.

The tourist industry of mayanization has reached a level in which we can hardly refer to the Maya Forest as a "periphery", subject as it is to heavy development, appropriation, and industrialization, as illustrated in the case of the Maya Train in Mexico. Given the globalized and internationalized territorial transformations in the Maya Forest – including tourism – the peripherality of the region, with its competing regionalizations and eco-regionalizations, may be challenged. At the same time, the developments of the Maya Forest incite questions about the human–nature divide within Borderlands Studies. Conservation work and protected areas have penetrated the supposedly uninhabited forests, resulting in mappings and counter-mappings by Indigenous people and environmentalists that are no longer characterized by empty spaces, as it was the case for historical maps attempting to draw political boundaries (e.g., Castillo et al., 2006). This challenges the perception of sedentary borderlands defined only by settled populations. Rather, as is the case of the Columbia River Forest Reserve example, borderlands are formed via identification and protection of hunting grounds and the existence of species other than humans, as well as historical mobilities.

The chapter also invites further research. One such topic (only implicitly present in this chapter) is related to the entanglement of geo- and biopolitics with the Mayas, which could be deepened in many ways. Another one is the necessity of deeper research about the historical role of the *Mayan rivers* that, as explained by the archaeologists cited in this chapter, may have contributed or not to the forest people' mobilities along the Maya Forest waterlands in a current context of their integration into the Mayan touristic narrative and development. The territory is vast, with many cases and contexts to be examined and understood. While the concept of the *Maya* continue to obtain solicited and unsolicited attention in the Maya Forest, perhaps the next Borderlands Studies will also explore what has been left underfoot of these – for example, the Garifuna people and the Black Atlantic, or the new settlers and migrants in the Maya Forest.

References

Aguilar Palafox, L. P. (n.d.). Regiones de Tabasco. DeTabascosoy. https://detabascosoy.com/regiones-de-tabasco/.

Almeida Poot, M., Jouault, S. & Rodríguez, Y. (2022). Las vías de la mayanización: Turismo, Tren Maya y representaciones de la mayanidad en la Península de Yucatán. *Maya America* 4:2, 62–88.

Anaya, J. (1998). Maya Aboriginal Land and Resource Rights and the Conflict over Logging in Southern Belize. *Yale Human Rights and Development Law Journal* 1, http://hdl.handle.net/20.500.13051/5708 (Consulted in March 2024).

Andersson R., Cothran B., Kekki S. (Eds, 2021). *Bridging Cultural Concepts of Nature: Indigenous People and Protected Spaces of Nature* (pp. 1–25). Helsinki University Press, Helsinki.

Arts, B., van Bommel, S., Ros-Tonen, M., Verschoor, G. (Eds, 2012). *Forest-People Interfaces: Understanding Community Forestry and Biocultural Diversity*. Wageningen Academic Publishers.

Bastos, S. & Camus, M. (2004). *El movimiento maya: una mirada en perspectiva*. FLACSO, Guatemala. Accessible: https://biblioteca-repositorio.clacso.edu.ar/bitstream/CLACSO/6312/1/movimiento.pdf (Consulted in March 2024).

Bastos, S., Cumes, A., & Lemus, L. (2007). *Mayanización y vida cotidiana. La ideología multicultural en la sociedad guatemalteca*. FLACSO-CIRMA-Cholsamaj, Guatemala.

Bengoa, José. (2000). *La Emergencia Indígena en América Latina* [*The indigenous emergence in Latin America*]. Fondo de Cultura Económica, Santiago, Chile.

Brysk, Alison. (2000). *From tribal village to global village: Indian rights and international relations in Latin America*. Stanford University Press, Stanford, CA.

Camargo, A. (2022). Land born of water: Property, stasis, and motion in the floodplains of northern Colombia. *Geoforum*, 131, 223–231. Cano, I. J. (2024). Leer el 'desorden'. Cambio agrario, campesinados y el Sembrando Vida. *Estudios Sociológicos*, 42(124), 1–17.

Canter, R. L. (2007). Rivers among the Ruins: The Usumacinta. *PARI Journal*, 7(3), 1–24.

Canter, R. L., & Pentecost, D. (2007). Rocks, Ropes, and Maya Boats; Stone Bollards at Ancient Waterfronts Along the Rio Usumacinta: Yaxchilan, Mexico to El Porvenir, Guatemala. *PARI Journal*, 8(3), 1–14.

Castillo, M., Toussaint, M. & Vázquez, M. (2006). *Espacios Diversos, Historia en Común*. Secretaría de Relaciones Exteriores.

Ceceña, A. & Prieto, S. (2023). *Mirando al Sur: Megaproyectos, fronteras e (in) movilidades. In Martínez, E., Gasparello, G. & Díaz, M. (Eds.). Territorios mayas en el paso del tren: Riesgos previsibles y posturas independientes sobre el Tren Maya* (pp. 25–51). Bajo Tierra Ediciones, Mexico City.

Colchester, M. (2003). *Salvaging Nature: Indigenous Peoples, Protected Areas and Biodiversity Conservation*. World Rainforest Movement and Forest Peoples Programme, Montevideo.

Convention on Biological Diversity. (1992). United Nations. https://www.cbd.int/ (Consulted in March 2024).

Convention on World Heritage, and lists of world heritage sites. (1972). United Nations. https://whc.unesco.org/en/convention/ (Consulted in March 2024).

de Vos, J. (2002). *Una tierra para sembrar sueños: Historia reciente de la Selva Lacandona, 1950–2000*. CIESAS, Mexico City.

Dorman, S. & García, Z. (2021). Panama's Supreme Court recognizes Indigenous Peoples' land rights and role as guardians of the environment. Blog-post. Center for International Environmental Law. https://www.ciel.org/panamas-supreme-court-recognizes-indigenous-peoples-land-rights-and-role-as-guardians-of-the-environment/ (Consulted in March 2024).

ECOSUR. (1995). *Protected Areas and Archaeological Sites – Mapping and database development to support conservation planning in the Selva Maya tri-national region*. Library, San Cristóbal de las Casas. El Colegio de la Frontera Sur, México.

Engle, K. (2010). *The Elusive Promise of Indigenous Development: Rights, Culture, Strategy*. Duke University Press.

Escalona Victoria, J.L. (2017). La manufactura de los mayas. Los orígenes de una epistemología política del presente. In Agudo Sánchiz, A., Estrada SaavedraM., BraigM. (Eds). *Estatalidades y soberanías disputadas: la reorganización contemporánea de lo político en América Latina* (pp. 117–142). El Colegio de México.

Escalona Victoria, J.L. (2018). Encapsulated history. Evon Vogt and the Anthropological Making of the Maya. In López Caballero, P., Acevedo-Rodríguez, A. (Eds). *Beyond Alterity: Destabilizing the Indigenous Other in Mexico* (pp. 244–262). The University of Arizona Press.

EU Delegation to Jamaica, Belize, the Bahamas, Turks and Caicos Islands, and Cayman Islands. (2023). Survaying their way to land rights: Belize Maya Women's GPS Mapping Team. Press release 06.12.2023. https://www.eeas.europa.eu/delegations/jamaica/surveying-their-way-land-rights-belize%E2%80%99s-maya-women%E2%80%99s-gps-mapping-team_en (Consulted in March 2023).

Fedick, S. (2003). In Search of the Maya Forest. In Slater, C. (Ed). *In Search of the Rain Forest* (pp. 133–167). Duke University Press.

Fisher, C. & Chase, A. (Eds, 2021). *Maya Anthropological Archaeology*. MDPI Books.

Ford, A. & Nigh, R. (2015). *The Maya Forest Garden: Eight millennia of sustainable cultivation of the tropical woodlands*. Routledge, New York.

Global Atlas of Environmental Justice. https://ejatlas.org/ (Consulted in March 2024).

Golden, C., & Scherer, A.K. (2013). All of a Piece: The Politics of Growth and Collapse in Classic Maya Kingdoms. *Archaeology*, 4, 157–171.

Graeber, D. & Wengrow, D. (2021). *The Dawn of Everything: A New History of Humanity*. Allen Lane.

Grandia, L. (2012). *Enclosed: Conservation, Cattle and Commerce among the Q'eqchi' Maya Lowlanders*. University of Washington Press.

Holmes, C. (2016). *Ecological Borderlands: Body, Nature, and Spirit in Chicana Feminism*. University of Illinois Press, Chicago.

Juniper, T. (2018). *Rainforest: Dispatches from Earth's Most Vital Frontlines*. Island Press.

Kauffer, E. F. (2019). From the abundance of waters to the scarcity of studies: Contemplating hydropolitics in Mexico-Guatemala and Mexico-Belize borders. In Guerrero García Rojas, H.R., *Water Policy in Mexico: Economic, Institutional and Environmental Considerations* (pp. 237–262). Springer.

Kauffer, E. (2023). Negotiating smallness in three regional contexts: Belize within Central America, the Caribbean, and neighboring Mexico. In Kolnberger T., Koff H. (Eds). *Agency, Security and Governance of Small States* (pp. 104–117). Routledge.

Laako H. (2016). Decolonizing Vision on Borderlands: The Mexican Southern Borderlands in Critical Review. *Globalizations* 13 (2), 173–187. <https://www.tandfonline.com/doi/abs/10.1080/14747731.2015.1076986>.

Laako, H. & Kauffer, E. (2021). Conservation in the Frontier: Negotiating Ownerships of Nature at the Mexican Southern Border. *Journal of Latin American Geography* 20 (3), 40–69.

Laako, H. & Kauffer, E. (2022). Between Colonizing Waters and Extracting Forest Fronts: Entangled Eco-frontiers in the Usumacinta River Basin. *Political Geography* 96. 102566.

Laako, H., Pliego, E., Ramos, D. & Marquez, B. (2022). International Relations, Challenged: Transboundary Conservation and Nature States in the Maya Forest. *Globalizations* 19 (8), 1288–1310.

Lakomäki, S. (2015). *Gathering together: The Shawnee people through diaspora and nationhood, 1600–1870*. Yale University Press, New Haven.

Lamb, D., & Lamb, G. (1954). *En busca de la ciudad perdida*. Noguer, Barcelona.

Lariagon, R. (2023). *Mayapolis*. Documentary film accessible: https://www.youtube.com/@mayapolis1627 (Consulted in March 2024).

Maler, T. (1901–03). *Researches in the central portion of the Usumacinta Valley; report of explorations for the Museum, 1898–1900*, Peabody Museum of American Archaeology and Ethnology.

Martínez Romero, E., Gasparello, G. & Díaz, M. (Eds, 2023). *Territorios mayas en el paso del tren: Riesgos previsibles y posturas independientes sobre el Tren Maya*. Bajo Tierra Ediciones, Mexico City.

Meerman, J. (2004). *Rapid Ecological Assessment: Columbia River Forest Reserve past Hurricane Iris*. A study commissioned by the Ya'axche' Conservation Trust and Toledo Institute for Development and Environment.

Miller, B & Miller, C. (1990). *Special Report: The Chiquibul Forest Reserve and the Considerations on the Establishment of a Biosphere Reserve for Belize*. Prepared for Minister Daniel Silva. Wildlife Conservation International, a division of New York Zoological Society. Belize Archives and Records Service.

Morelet, P. (1857/2015). *Viaje a América Central, isla de Cuba y Yucatán*. Universidad Nacional Autónoma de México.

Nations, J. (2006). *The Maya tropical forest: People, parks, and ancient cities*. University of Texas Press, Austin.

Nigh, R. (2002). Maya Medicine in the Biological Gaze: Bioprospecting Research as Herbal Fetishism. *Current Anthropology* 43:3, 451–477.

Obregón Rodríguez, C., & Liendo Stuardo, R. (2016). *Los antiguos reinos mayas del Usumacinta Yaxchilán, Bonampak y Piedras Negras.* Fondo de Cultura Económica, El Colegio de México.

Peréz Ruiz, L. (2018). La lucha del pueblo maya por el territorio y los recursos bioculturales. *Amérique Latine: Histoire & Memoire, Les Cahiers ALHIM online* 36. https://doi.org/10.4000/alhim.6963.

Pieck, S. & Moog, S. (2009). Competing entanglements in the struggle to save the Amazon: The shifting terrain of transnational civil society. *Political Geography* 28:7, 416–425.

Primack, R., Bray D., Galletti H., & Ponciano I. (Eds, 1998). *Timber, tourists, and temples: Conservation and development in the Maya Forest of Belize, Guatemala and Mexico.* Island Press, Washington.

Ramsar Convention on Wetlands. (1971). https://www.ramsar.org/ (Consulted in March 2024).

Readman, P., Radding, C. & Bryant, C. (Eds, 2014). *Borderlands in World History 1700–1914.* Palgrave Macmillan.

Restall, M. & Solari, A. (2020). *The Maya: A Very Short Introduction.* Oxford University Press, New York.

Richards, S. & Kruger, A. (2023). El Tren y la Peninsula. Produced by Dos Tohs Films and Mutual Aid Media. Documentary film accessible at Mirada Nativa: https://www.miradanativa.org/blog/estreno-internacional-en-miradanativa-de-el-tren-y-la-peninsula/ (Accessed in March 2024).

Ruz, M.H. (2010). *Paisajes de río, ríos de paisaje. Navegaciones por el Usumacinta.* UNAM, IIFL, Centro Peninsular en Humanidades y Ciencias Sociales/Consejo de Ciencia y Tecnología del Estado de Tabasco, México.

Scherer, A.K., Golden, C., Houston, S., Matsumoto, M.E., Firpi, O.A.A., Schroder, W., … & Hruby, Z.X. (2022). Chronology and the evidence for war in the ancient Maya kingdom of Piedras Negras. *Journal of anthropological archaeology*, 66, 101408.

Scherer, A.K., & Golden, C. (2012). *Revisiting Maler's Usumacinta: recent archaeological investigations in Chiapas.* Precolumbia Mesoweb Press.

Scott, J. (1990). *Domination and the Arts of Resistance: Hidden Transcripts.* Yale University Press.

Schroder, W., Golden, C.W., Scherer, A., Álvarez, S.D.P.J., Dobereiner, J., & Cab, A. M. (2017). At the crossroads of kingdoms: Recent investigations on the periphery of Piedras Negras and its neighbors. *PARI Journal*, 17(4), 1–15.

Períodico Oficial, 29 de junio. (1994). . http://periodicos.tabasco.gob.mx/media/1994/88.pdf. Secretaría de Gobierno.

Slater, C. (Ed, 2003). *In Search of the Rain Forest.* Duke University Press.

Solano, X.L., Leyva, X., Alonso, J., Hernández, R.A., Escobar, A., Köhler, A., Cumes, A., Sandoval, R., Speed, S., Blaser, M., Krotz, E., Piñacué, S., Nahuelpan, H., Macleod, M., Intzín, J.L., García, J.L., Báez, M., Bolaños, G., Restrepo, E., … de Sousa Santos, B. et al. (2015). *Prácticas otras de conocimiento(s): Entre crisis, entre guerras.* CLACSO.

Stephens, J.L., (1843). *Incidents of Travel in Yucatan*: (Vol. 1). Harper & Brothers.

Stephens, J.L., & Catherwood, F. (1841). *Incidents of Travel in Central America, Chiapas and Yucatan… With Illustrations by Frederick Catherwood, Etc.* Dover Publications.

Stevens, S. (2014). Indigenous Peoples, Biocultural Diversity and Protected Areas. In Stevens, S. (Ed). *Indigenous Peoples, National Parks and Protected Areas: A New*

Paradigm Linking Conservation, Culture and Rights (pp. 15–46). The University of Arizona Press.

Takeshita, C. (2000). *Bioprospecting and Indigenous Peoples' Resistances, Peace Review*, 12:4, 555–562, doi:10.1080/10402650020014645.

Tarrow, S. (1998). *Power in Movement: Social Movements and Contentious Politics*. Cambridge University Press.

Toledo Maya Cultural Council and the Maya People of Southern Belize. (1997). *Maya Atlas: The Struggle to Preserve Maya Land in Southern Belize*. North Atlantic Books.

Torres-Mazuera, G. (2023). Dispossession through land titling: Legal loopholes and shadow procedures to urbanized forestlands in the Yucatán Peninsula. *Journal of Agrarian Change*, 23(2), 346–364.

Tuck, E. & Yang, K. (2012). Decolonization is not a metaphor. *Decolonization: Indigeneity, Education and Society* 1:1, 1–40.

Turnbull, C. (1961). *The Forest People: A Study of the Pygmies of the Congo*. Simon and Schuster.

UNESCO MAB Website (2024). Man and Biosphere Programme MAB: https://www.unesco.org/en/mab (consulted in August 2024).

Wainwright, J. (2008). *Decolonizing Development: Colonial Power and the Maya*. Blackwell Publishing.

Ybarra, M. (2018). *Green wars: Conservation and decolonization in the Maya Forest*. University of California Press, Oakland.

3 Rethinking Transboundarities

Connectivities of Water and Conservation in the Maya Forest Waterlands

Edith Kauffer and Hanna Laako

Introduction

As dynamic borderlands, the Maya Forest Waterlands face deep transboundary water and forest issues. Forests extend upon borders and waters cross political boundaries; thus, the Maya borderlands are mainly forest waterlands filled by forests and waters. Nevertheless, deforestation is continuing throughout the region; water resources show a growing imbalance, and the area is witnessing a tendency toward extreme events in the context of climate change – droughts, floods, and high temperatures. This chapter contributes to understanding the transboundary angle of forest waterlands, focusing on the dynamics established by the presence of political boundaries and geopolitics, but also by the local realities of the riparian population. It explores the dynamics of the connectivities related to the transboundarities of water and conservation, which is based on the entangled presence of waters and forests in a transboundary context.

The concept of *transboundarity* has not been commonly discussed by the English international academic literature. Only a few papers on transboundary waters and river basins have mentioned the notion, though without proposing a clear definition. For example, Keskinen et al. (2016, p. 15), in their research on three transboundary river basins, state that *transboundary* refers to activities or issues that are "connected in multiple ways and are therefore transboundary in multiple ways". This perspective only establishes a multiplicity of connections associated with the concept. Focusing on the water–energy–food nexus, the authors evidence scales, key players, and political aspects that broaden and foster cross-border cooperation as components of transboundarities about water analyzed through connections in the shared river basins they have studied.

Transboundarity is also referred to in international reports on water issues as a common notion or evidence (e.g., Niang-Diop et al., 2002). It generally focuses on three main aspects: the existence of natural resources or geographical conditions that transcend political boundaries; the presence of a legal framework for cooperation generally focusing on states; and the different kinds of relations or movements that cross international boundaries, such as timber logging or local resources sharing in the case of the Maya Forest Waterlands. More precisely, the scarce literature has mentioned the following main

DOI: 10.4324/9781003429050-4

examples of transboundarities: natural transboundarities regarding shared resources such as fish, international watercourse issues and law in this field to promote cooperation, and the question of protected areas located in borderlands (Keskinen et al., 2016; Niang-Diop et al., 2002).

As international organizations have promoted transboundary conservation at least since the 1990s, the concept of transboundarity has been mainly addressed in terms of safeguarding biodiversity beyond and along political borders as well as ecosystem connectivity between countries. Given that this notion is also at the core of the creation of the concept of the Maya Forest, we critically address it in this chapter. In this context, Laako et al. (2022, p. 15) have shown that in terms of transboundary conservation, the conceptualization of transboundary tends to include phenomena and aims that extend both "across and along borders". In the same vein, following Iglesias-Prieto (2021, p. 114), it is essential to consider

> … that transboundarity involves much more complex processes, perspectives, subjects and identities that are commonly experienced as processes of lesser to greater integration. Nonetheless, transboundarity […] does imply a much more complex practice of citizenship by participating simultaneously in multiple social, cultural, political and, of course, geographic spaces. Evidently, transboundarity functions beyond national categories.
>
> Own Translation from Spanish

As a matter of fact, transboundarities are thus defined both according to the ontological existence of borders and to the human and nonhuman capabilities to extend beyond political boundaries, creating connections. Nevertheless, these definitions are predominantly focused on political or jurisdictional borders, which are often considered fixed, although they are often formed in natural environments or along natural elements making access difficult to humans, such as rivers, which are not stable but fluid, as this chapter will show in the case of the Maya Forest.

To further deepen the reflection on transboundarities, scholarship in Spanish has focused on borders and borderlands and referred to the concept of "transboundaries" (*transfronteras*) as follows, starting from a reflection on the meaning of the *trans* prefix:

> Borders cannot escape the prefix trans that defines them as transitive spaces, transshipments, transports, transfers, crooked deals (*transas* in Spanish) and transactions. The Latin prefix *trans* is an inherent part of borders; it delimits them, summons the other side and incorporates it denoting changes and moves. Trans refers to the condition of locating oneself through but also beyond. Along with the traditional views of borders that emphasize confines, limits, what is located in front, borderland, doors, entrances, and thresholds, transboundaries incorporate the other side, the beyond, change and relocation as inseparable aspects. Transboundaries is a concept that conjugates the oxymoron mentioned above;

> they are spaces that refuse to only deal with one of the conditions or sides that integrate it. Transboundaries are not exhausted in trenches, nor in impassable limits, nor in innocent spaces, nor in illusions that can be dismantled at will, nor in sites of horizontality, nor in mere encounters, nor in membranes, nor in the porous condition of spaces. Thinking from the transboundary implies incorporating complex geopolitical, social, and cultural conditions.
>
> Valenzuela 2021, p. 9–10, Own Translation from Spanish

Transboundarity is thus derived from the notion of transboundary – as a noun, as used in the preceding paragraphs, that names processes, spaces, and relations related to one or more transcended or transgressed boundaries, or that invites to include, analyze, and consider the connections beyond the boundary. As proposed by Valenzuela (2021, p. 15), "the notion of transboundaries (*transfronteras*) refers to areas defined by bordering, transboundary coexistence, the conjunction and disjunction of processes, the injection of power that commands, imposes, injects, as well as the presence of connective processes, contact zones, transboundary areas" (own translation from Spanish). Nevertheless, as cited above, he also regards transboundarity as limited or reduced to "geopolitical, social and cultural conditions", which do not include environmental aspects fundamental to transboundary conservation projects such as that of the Maya Forest. Political borders are a human creation and, therefore, human-centered. It is thus understandable that social and political scientists have predominantly interpreted transboundarities from an anthropocentric viewpoint. This human-centeredness in transboundarities, however, creates a conceptual dilemma and misinterpretation when employed in such endeavors as transboundary conservation, which are not primarily concerned with human/political/social transboundarities but with those of ecosystems and species.

Pointing out the vagueness of "transboundarity" in terms of political or jurisdictional and social borders has been adequate. In this chapter, we suggest a missing link: species connectivity as transboundarity, which tends to be the primary motif of transboundary conservation, rather than that of humans divided by bordered jurisdictions. In this chapter, we rethink transboundarities from the perspective of forest waterlands, which is another kind of transboundarity between ecosystems and across many different borders, and emphasize the different ways in which these may be understood – particularly beyond our own field of political science, which has been predominantly focused on political borders. Additionally, we show that the political borders in the Maya Forest involve rivers and waterlands that can hardly be considered fixed categories but instead fluid borders – in a way transboundary themselves – and ecosystemic borderlands (forest waterlands), which contain high biodiversity.

To continue, we address a similar challenge related to transboundarities within conservation: those between ecosystems. Conservationists have introduced the concept of connectivity beyond political borders to promote transboundary conservation related to ecosystems and species. However, the understanding of "transboundarities" and borderlands between ecosystems is slowly emerging. We shed light on some of those questions related not only to transboundarities along and across political borders, but also along and across ecosystems in the Maya Forest.

In this chapter, we focus on waterlands as a hidden issue within Maya Forest conservation, which has so far emphasized forests. Nevertheless, the water–land interface is key to understanding the Maya Forest waterlands as coined by ecologists (Talley et al., 2006), and tropical rivers are especially characterized by the seasonality (Syvitsi et al., 2014) that defines the changing shapes of waterlands. In 2021, the Food and Agriculture Organization of the United Nations (FAO) published a report exploring water–forest management (FAO et al., 2021), which stated that forest and mountain ecosystems are the source of more than 75% of renewable water supply. While there are some individual studies on forest–water relationships, this recent report is the first global publication on the monitoring, management, and valuing of forest–water interactions. The main forest–water ecosystems mentioned are mangroves, peatland forests, tropical montane cloud forests, and dryland forests. While the importance of forest–water management was first recognized in the Shiga Declaration on Forests and Water in 2001 and some thematic studies were conducted by the FAO in 2008, most scientific and inventory work exploring this particular nexus dates from the past decade. The report emphasizes that, more often than not, forest–water interfaces are taken for granted as a default byproduct, whereby it is assumed that forest management will automatically improve water supply and vice versa.

The concept of connectivity is intimately related to safeguarding biodiversity and transboundary conservation. It was developed by landscape ecologists in the 1980s, when isolation was addressed as an element hindering movement among resource patches (Crooks & Sanjayan 2006). Connectivity became a key feature of metapopulations. While the term continues to be debated it has contributed to understanding accelerated habitat fragmentation. In the 1990s, the connectivity trend extended to corridors as a conservation tool, as it was understood that isolation prevented or eroded biodiversity; closed and isolated protected areas, for example, could not conserve many migratory species. Habitat destruction and fragmentation have been recognized as prime threats to (native) biodiversity. Thus, conservation tools have increasingly addressed connectivity, referring to the movement of organisms or processes and including the idea that more movement results in more connectivity (Crooks & Sanjayan, 2006). While biodiversity conservation focuses on species connectivity, movement in nature also involves elements such as soil, fire, wind and water, plants and animals, ecological interactions, ecosystem processes, and natural disturbances. Therefore, since the 1990s and together with transboundary conservation, connectivity has evolved as a strategy focused on creating corridors both between protected areas and more broadly in a landscape extending beyond them and in between them. In other words, "transboundary" is not meant to focus on human or jurisdictional extension over borders, but on those of ecosystems and species. In our view, beyond isolated units and enclosures created politically but not reduced to human-centered aims, biodiversity connectivity is a key component of transboundarities analyzed from a forest waterlands perspective.

This chapter focuses on the realities of the Maya Forest waterlands in terms of transboundary relations regarding water issues and conservation. The natural transboundary conditions of forests and waters tend to be in contrast with the political realities of their management imposed by international and other types of boundaries. The chapter considers the roles of diverse actors as well as the dimensions of space and time, offering a distinct and complementary perspective to that of Chapter 4, which is centered on transboundary river basins.

The first part of the chapter seeks to characterize the Maya Forest waterlands in terms of the diversity of transboundary waters, the meanings of boundaries for waters and forests, and their significance in terms of transboundarities. Crossing boundaries and categories of waters (Kauffer, 2020), we define and illustrate, for different points of the region, the notions of transboundary rivers, international rivers and sources, transboundary water bodies, transboundary, international, and shared river basins, as well as transboundary wetlands. This characterization is bound by international law combined with political outcomes and management. Regarding conservation issues, this chapter also describes the existing categories of transboundary conservation at an international scale and their concrete expressions in the Maya Forest waterlands. Finally, it explores the possible connected and disconnected entanglements of water and forests. The section highlights that the connections between tropical rivers and borders are complex and diverse, and that the latter are more fluid than fixed along the analyzed waterlands.

The second part of the chapter focuses on transboundary conservation and explores how water is involved (or not) in the dynamics that have so far been focused on forests. The Maya Forest elucidates a space for terrestrial ecosystems and archaeologists, as shown in Chapter 2. The Maya Forest forms part of the strategy of transboundary conservation, which comprises a broad network of internationally adjoining protected areas – IAPAs (e.g., Laako et al., 2022). The Maya Forest was recognized as a strategic complex for transboundary conservation, for example, in Mittermeier et al. (2005, p. 147–157), which presented 28 such areas around the world. These 28 areas were predominantly forests, although some waterlands do appear in the names, especially in the Americas, such as the *Laguna Madre*, characterized as the major transboundary wetland on the Texas–Tamaulipas border (Mittermeier, 2005, p. 24–25). The section on the "Maya Tropical Forest", elaborated by Nations et al. (2005, p. 146–157), included descriptions of the protected areas, main actors (such as the international conservation organizations, local NGOs and government institutions), state of the forests, and Mayan history. However, seasonally flooded forests, the lowland swamp forests of Tabasco, and the "largest freshwater wetland in Central America" were also mentioned (Nations et al., 2005, p. 147). Nations et al. (2005) considered that the most promising areas for conservation included the Bladen Nature Reserve and

Columbia River Forest Reserve in Belize, the Tzendales region and Lacantún Biosphere Reserve in Chiapas, and the Mirador-Calakmul region in the Petén-Campeche borderlands. Prior to this, active ecoregional mapping of the Maya Forest had taken place, as explained in Chapter 1 (e.g., García & Secaira, 2006). Laako et al. (2022) traced the origins of the Maya Forest concept as part of the transboundary conservation strategy, locating earlier mappings from 1995 (ECOSUR, 1995). These maps include the Maya Forest hydrography. Yet, the main areas are forests and protected areas, excluding, for example, marine conservation areas and most of the riverlands of Tabasco.

Many developments have taken place within the Maya Forest since 1995. Although the transboundary efforts in the Maya Forest have continued to predominantly focus on adjoining protected areas and their patrolling, fieldwork results suggest a growing emphasis on waters –particularly by the smaller NGOs. As shown in this part of the chapter, a shift toward "waterlands" as well as transboundary wildlife conservation efforts is taking place, particularly in terms of monitoring waterholes for the endangered fauna. Simultaneously, however, the Maya Forest conservation is challenged by transboundarities related to ecosystems. Marine and forest ecosystems are often separated as categories, and only recently studies have emerged that explore the transboundarity between these ecosystems as areas of potentially high biodiversity.

From the concept of transboundaries just evoked, which suggests connectivity but also integration, we propose a twofold perspective, which deals with the diverse forms of *trans*gressing the political borders in the Maya Forest waterlands – one starting with waters and the other one with forests. The first one deals with the analysis of transboundary waterlands and the diverse entanglements between waters, forest, and borders. The second discusses transboundary conservation in borderlands and its relations with water.

Transboundary Forest Waterlands: Waters, Forest, and Entangled Borders

All over the studied Maya Forest region, forests and waters are entangled but also crossed by different types of boundaries. This section proposes to define and characterize these entanglements, intersecting first the boundaries and categories of waters (Kauffer, 2020) and explaining the continuity of forests across borders. Transboundary waters and forests show substantial differences owing to the mobility and fluidity of waters, which "naturally" flow across human boundaries, including international borders, whereas forests are immobile and rooted in the land.

Forty-five years ago, John Waterbury (1979, p. 2) stated about the Nile River that "rivers have a perverse habit of wandering across borders [...] and nations states have a perverse habit of treating whatever portions of them that flows within their borders as a national resource of its sovereign disposal". Unlike forests rooted in the soil, the sometimes-uncontrollable movement of waters is conceived as a threat by states to their sovereignty. Despite water being controlled by dams and other infrastructures, such as water transfers between river basins, surface – and, above all, subterranean waters – flow and cross boundaries without major limitations. The issue of transboundary waters has been so complex and challenging for states that the 1997 New York Convention on the Law of Non-Navigational Uses of International Watercourses was only ratified in 2014, 17 years after being adopted by the United Nations General Assembly, owing to a reluctance of states to adhere to it. It was extremely difficult to reach the number of 35 states required for this convention to enter into force. Among the three states that share waters in the Maya Forest waterlands, Mexico voted in favor of the United Nations' resolution in 1997; Guatemala abstained, and Belize was absent. By 2024, none of the three states has ratified the New York Convention.

The transboundary dimensions of waters are potentially disruptive and, as suggested by diverse authors, "emotional" or subject to affective responses and points of views at collective or individual scales (Seide & Fantini, 2023; Sehring & Wolf, 2023). During fieldwork in the region between 2003 and 2024 around and about transboundary waters, state authorities, particularly the foreign ministries of Mexico and Guatemala and the Water Authority of Belize, have prohibited our research teams and other colleagues from even using the notion of "transboundary" in association with waters during meetings, academic events, and face-to-face interviews.

What are the transboundary waters of the Maya Forest waterlands and how are they entangled with borders? In addition to Chapter 4, which proposes an approach based on transboundary river basins, we focus on rivers – transboundary and international ones – and on water bodies to further analyze specific waterland corridors as expressions of transboundarities following the definition proposed in Chapter 1. Each section presents a case that refers to a specific waterland with a particular convergence of rivers and borders. By considering the connections with forests in the Maya waterlands, we first analyze how waters cross borders; second, how they merge; third, the state of the border when the river disappears; and finally, what happens when international rivers coincide with wetlands and the implications for both borders and rivers.

When Waters Cross Political Borders: Transboundary Rivers in the Maya Forest Waterlands

If we refer to international law combined with political issues, transboundary waters include diverse water bodies and, above all, their entanglements with boundaries. Boundaries may be diverse: political, social, cultural, or ecological (Kauffer, 2024). For example, river basin limits cross Indigenous territories, which are also entangled with municipal delineations. None of them coincide on the map, and they also correspond to different local perspectives *in situ*. The present description only deals with international political borders and waters without including other types of boundaries.

The first component of the entangled waters and boundaries are transboundary rivers or sources, which we propose to define as streams crossing an international border without delineating it, flowing from an upstream to a downstream location. To understand how this first category of transboundarities is present all over the Maya Forest waterlands, we must consider each international border separately.

The studied region is traversed by three international borders: Guatemala–Mexico, Belize–Mexico, and Guatemala–Belize. The first one was established in 1882 following decades of conflicts between Guatemala and Mexico. As the longest international border among the three mentioned, it extends for 986 km, although we must mention that its length depends on the consulted source (Kauffer Michel, 2013). Many transboundary rivers that cross this border and those in the Maya Forest waterlands generally flow from Guatemala – located upstream – to Mexico – located downstream. Transboundary rivers generally change their name when they cross an international border, and the toponymy corresponds to local denominations, topography, and hydrography as well as nationalist expressions. This is not a peculiarity of the Maya Forest waterlands but a characteristic of transboundary rivers between Mexico and Central America, which can be observed all over the world. Although they play a role as shared and connecting elements, these transboundary rivers also define diverse perspectives of the political boundaries and realities located beyond the border – in other words, they also favor a kind of social and political connectivity.

Several transboundary streams cross the international border between Guatemala and Mexico. Some of them are major rivers, such as the Candelaria River or the San Pedro Mártir/ San Pedro Missicab River, while others are upstream tributaries of the Usumacinta River (Xalbal/Chajul River and Ixcán River). Kauffer Michel (2017, p. 74) illustrated this situation by describing a 30-kilometer borderland characterized by water borders with multiple streams crossing the border line – a profusion of rivers which forms a "water web", "water border", or 60 km of "water borderland".

The second border separates Mexico and Belize and was established in 1893 between Mexico and British Honduras, the former colony. Belize became independent in 1981, decades after the signature of the Boundary Convention. Nevertheless, it is still in force and has not been contested by either state. The land border only comprises small streams but no major transboundary rivers, as the whole border is mainly delineated by the Hondo River, which forms an international river.

The third border is a complex issue. It could be considered a non-formal border from a legal perspective, as no border treaty has been concluded between Belize and Guatemala as independent states. Nevertheless, according to Belize, the contemporary border has been defined in the Treaty of 1859 between British Honduras, the former colony, and Guatemala. However, Guatemala argues that the treaty is invalid owing to Great Britain's noncompliance with the stipulation in the treaty that it should build a road from the border to the Caribbean Sea. For this reason, Guatemala only refers to a so-called "adjacency line". Consequently, this "border" has given rise to a deep and complex conflict apparently focused on the demarcation but which, above all, stands for Guatemala's claims to parts of Belize's territory. Guatemala delayed the recognition of Belize's sovereignty after its independence in 1981 for 10 years, and the tourist maps at the beginning of the 1990s still showed the whole of Belize inside Guatemalan territory. During fieldwork at the Ceibo border crossing in 2022, we saw such a map hanging on the wall of the Guatemalan customs office as recently as then.

The territorial conflict that affects more than the half of Belize's territory also includes a maritime border dispute and claims to islands in the Caribbean Sea. In June 2019, the case was brought to the International Court of Justice, whose jurisdiction was accepted by both states after a nationwide referendum in each country in 2018 and 2019. Today, skirmishes at the land and fluvial border (Sarstoon River, which is international) are common (Romero & Cegarra, 2024), with their intensity occasionally increasing. The Guatemalan Army often patrols near the adjacency line in a region characterized by numerous smuggling operations. The Sarstoon River forms the southern line between Guatemala and Belize. The terrestrial line – with Guatemala to the west and Belize to the east – between both countries is crossed by three major Belizean rivers that are transboundary with Guatemala: the Belize (the most important one), Moho, and Temash Rivers. Out of 16 major watersheds, Belize shares five with its neighbors. The Hondo River basin – also an international river – is shared with Mexico, and four rivers are shared with Guatemala, with the international Sarstoon River added to the three mentioned above. All of them flow into the Caribbean Sea as, in the case of the three major transboundary rivers, Guatemala is located upstream and Belize downstream.

In this chapter, we use "border" when referring to Belize's perspective and "adjacency line" when speaking from the Guatemalan's point of view. This is not a position statement; we only take into account two interpretations of the same reality, which also express contrasting political positions regarding its delimitations.

When Rivers Are Borders and Borders Merge into Rivers: International Rivers at the Lacandon Forest Waterlands and Laguna del Tigre-Gulf Riverlands

Beyond transboundary rivers, the entanglements of boundaries and waters in the Maya Forest waterlands also reveal the existence of international rivers. As defined by international law, international rivers delineate international borders. The Maya Forest waterlands are crossed by international rivers of different length and flow, which shape specific borderlands in terms of politics and geopolitics and are fully intertwined with our concept of waterlands. Additionally, international rivers raise the question of the clear location of the international border in tropical regions. At the international scale, there are three solutions to define a fluvial border. The first one is to locate the border on one of the riverbanks, which implies exclusive sovereignty of one of the states over the river, such as in the case of the San Juan River between Nicaragua and Costa Rica. The border is located on the Costa Rican riverside; thus, the river belongs to Nicaragua. This is the most contested international river and basin in Central America with a long history of disputes and appeals to the International Court of Justice (ICJ) (Boeglin, 2013; Kauffer Michel, 2018). This option usually generates conflicts and complex transboundary relations owing to the impossibility of the river being equitably shared. In the second solution, the border is located in the middle of the river, which technically signifies the middle line of the water mirror. The third solution, inspired by the large European rivers that were navigable in the nineteenth century, places the border at the thalweg, which is a German notion defining the deepest channel. This solution tries to guarantee an equitable use of the river for navigation. The last two options are complex when dealing with tropical rivers characterized by varying flows all over the year, as both the middle of the stream and the thalweg change position according to the river flow. Nevertheless, they supposedly favor peaceful transboundarities based on connectivity, owing to the underlying equity principle of the international boundary.

Between Guatemala and Mexico, the Usumacinta River delineates a 365-kilometer-stretch of the border that separates the Petén Department in Guatemala from the state of Chiapas in Mexico. The Usumacinta River is the largest river in Mexico and Central America and flows from the south (Guatemala) to the north (Mexico) to discharge into the Mexican Gulf. Using a Geographical Information System (GIS), Kauffer Michel (2013) determined the length of the river by measuring its major tributary in Guatemala, the Chixoy River. The Usumacinta's total length is 1,114 km, made up of 363 km upstream in Guatemala as the Negro and Chixoy Rivers, 365 km midstream, and 386 km downstream through Mexico. What is also interesting in this respect is the Usumacinta River's function of delineating the borders between federal states in Mexico. When the international fluvial border comes to an end, the Usumacinta River then divides the states of Tabasco and Chiapas, and along part of its route downstream, it separates Tabasco from Chiapas and Tabasco from Campeche. Consequently, the Usumacinta fulfils different bordering functions along its course, at international and subnational scales. In this case, transboundarities vary according to the location on the river and the type of boundary.

As an international river, the Usumacinta is located midstream in the whole river basin and extends both to the Lacandon Forest Waterlands and the Laguna del Tigre-Gulf Riverlands, as defined in the former chapters, although the two corridors are also entangled with each other. In this case, transboundarities are principally defined by the connectivity established by waters but also by the forests, rather than by a precise boundary. This is because, at this location, the waterlands are dominated by three confluences of major rivers. The first confluence is located in the south, where the Usumacinta River receives its major tributary from Guatemala, which also corresponds to its main upstream section, the Chixoy River. This confluence marks the beginning of the 365-kilometers-long fluvial international border. Further north, the Usumacinta River merges with another Guatemalan tributary, the La Pasión River. Eventually, it merges with the Lacantún River, a large river that flows from Mexico, also catching some of its upstream tributaries originating in Guatemala (Image 3.1).

The Lacandon Forest Waterlands and the neighboring Laguna del Tigre-Gulf Riverlands are characterized by a major confluence of watercourses around the great Usumacinta River. The south of this corridor is formed by an alluvial plain, where the meandering river circulates in a tight bed with a high risk of floods and has created several oxbow lakes. This tropical riverine landscape is unstable, as the river may rise by an average of 12 meters in some places. The waters of the Usumacinta rise and fall with the alternating dry and rainy seasons, and the riversides emerge and disappear depending on the water level. During extraordinary events in the rainy season, such as tropical storms or hurricanes upstream or midstream, the Usumacinta River becomes enormous, and at some confluences, the river overflows, dominates the land, and invades the forests. When the waters fall, the mix of sediments and eroded lands transported by the stream creates new landscapes, and young vegetation grows in places where some weeks before there was a sedimentary deposit (a river

Image 3.1 The Lacantún-Usumacinta Confluence
Source: Edith Kauffer, January 2018

beach) or, in some cases, cultivated land. As it occurred in November 2020 following the Eta and Iota storms, the river's course may change and create new riversides and interactions between water, land, and forest.

Archaeologists mention a "physical and cultural confluence that created a gathering of waters and communities" (Schroder et al., 2021, p. 690) in a region they call the Western Maya Lowlands. This floodplain, our Lacandon Forest Waterlands, which joins with the Laguna del Tigre-Gulf Riverlands, clearly illustrates the water–land nexus that questions the Western ontology separating water and land (Cortesi, 2016). Neither water nor land are static, but melt into each other: when water rises, land disappears, and when water diminishes, rocks, sediments, and fertile soils arise. The Lacandon Forest Waterlands, together with the Laguna del Tigre-Gulf Riverlands, shape a fluid landscape, a hybrid environment produced by the confluences of diverse rivers and by the ebb and flow of waters (depending on climate and seasons) but also by politics owing to the empirical reality of the fluvial border and the whole borderlands region. Throughout the alluvial plain, the Lacandon Forest Waterlands, together with the Laguna del Tigre-Gulf Riverlands, express the vibrancy of an aqueous landscape deeply influenced by the dynamism of the coalescing rivers. In this context of connectivity around rivers but also among forested lands and waters, transboundarities flourish according to local relationships and the specific characteristics of each fragment of waterlands.

From the archaeological site of Yaxchilán, situated on an omega-shaped peninsula on the Mexican riverside, the Usumacinta River moves north and enters a karstic crest. After the archaeological site of Piedras Negras located on the Guatemalan riverbank, large canyons and rapids render navigation difficult (Image 3.2). The international river border then gives way to the interstate border between the states of Chiapas and Tabasco at a place called The Line (*La Línea*, in Spanish). Moving further north, the Usumacinta River is entirely navigable downstream from Boca del Cerro – a canyon that also gives its name to a protected area in Tabasco, Mexico – down to its lower confluence with the Grijalva River and the San Pedrito River (an Usumacinta River downstream tributary), 361 km further downstream. This confluence called Tres Brazos (*Three Branches*), is situated 24 km from the rivers' joint discharge into the Gulf of Mexico (Image 3.3). From the confluence to the Gulf, the river is called Grijalva, although the major flow comes from the great Usumacinta River. In this case, the toponomy also expresses the Mexican state's perspective and politics that bestow major importance on the Grijalva River owing to the characteristics of the Grijalva watershed, which features four large dams, two capitals of federal states, and a more urbanized river basin.

In the alluvial plain that forms part of the Lacandon Forest Waterlands and the Laguna del Tigre-Gulf Riverlands, the rhythm of the Usumacinta River's movements through time sets the pace of the villages' daily life and the transboundary relations on both riversides: "a gathering of waters and

Image 3.2 Entering the Canyon and Rapids from the Piedras Negras Archaeological Site. Picture Taken from Guatemala with Mexico on the Opposite Riverside
Source: Edith Kauffer, January 2023

Image 3.3 Tres Brazos, Where the Usumacinta River (Left) Merges with the Grijalva River (Right) and the San Pedrito River (Middle)
Source: Edith Kauffer, March 2018

communities", according to an archaeologist (Schroder et al., 2021, p. 690), which is also a contemporary reality. As a world full of waters, it has been drained in the past to allow for agricultural activities and transformed to build reservoirs as a strategy to conserve water for the dry season. At this point, the fluidity that characterizes this landscape produced by the rivers and forests and by the transcendence of boundaries between water and land (Camargo, 2021) – the interface of water, land, and forest – is also crossed by the international border. All these elements contribute to connectivity and express transboundarities.

According to the border treaty between Mexico and Guatemala signed in 1882, the border is located at the thalweg (i.e., the deepest channel) following the nineteenth-century European tradition for navigable rivers. This kind of border definition creates difficulties for tropical rivers because of the large variations in river flows (Images 3.4 and 3.5). As stated for the Suchiate River, another international river between Mexico and Guatemala flowing into the Pacific Ocean, i.e., out of the Maya Forest waterlands (Kauffer, 2019), the thalweg as a definition of an international river promotes the border's mobility according to the changes in the river level.

Although only part of the Usumacinta's 265-kilometers-long international course is navigable, and only a limited stretch is occupied by villages on both sides as the Mexican riverside is more accessible by road than the Guatemalan one, both local populations have mentioned the issue of the moving fluvial

Image 3.4 High Water Levels at the Usumacinta River, with Guatemala on the Opposite Riverside (Same Location as the Following Picture)
Source: Edith Kauffer, January 2023

Image 3.5 Low Water Levels at the Usumacinta River, with Guatemala on the Opposite Riverside (Same Location as the Previous Picture)
Source: Edith Kauffer, August 2022

border during fieldwork. When we ask riverside communities for the location of the fluvial border, their interpretation of the demarcation line differs from that of the thalweg and generally locates the border on the middle line of the water mirror, which also moves according to the flows. This is a local interpretation of the border's location according to riparian needs and following the changing river flows. Although the exact location of the border seems to be clear for the riverside inhabitants, the situation is constantly changing throughout the year and over the years (Images 3.4 and 3.5).

According to a boatman who lives at and often crosses the confluence between the Lacantún and Usumacinta rivers, the borderline on the river is clear, although moving:

RESEARCHER: When the river grows more, does the line stay the same here or does it move?

BOATMAN: No, when the river grows, the river reaches there and always ... always according to the history, it must be the middle, the middle of the river.

RESEARCHER: The middle of the river, ok. So, does the line move a little bit? Because if the river gets bigger this way, and not so much that way ...

BOATMAN: And the line moves a little bit this way and when the river goes down again, then it moves that way again.

Fieldwork, January 20, 2018; Image 3.6

Image 3.6 The Lacantún-Usumacinta Confluence and the Fluvial Border: A Mexican Boat in the Foreground and a Guatemalan One in the Background
Source: Edith Kauffer, January 2018

During fieldwork, we have found that the border's empirical location also depends on local transboundary customs and practices aimed at guaranteeing the shared uses of the river for transport, which is possible from the confluence with the Chixoy River in the south to the archaeological site of Piedras Negras located upstream, at the entrance of the most dangerous rapids in the north (Image 3.2). Nevertheless, at each point of the border and riverside, we have encountered different perspectives according to the location of the line, depending on individual and collective experiences and on the transboundary interactions with the river, but also with natural surrounding resources, such as the forest. For Indigenous inhabitants working in a small biological station located to the north of the three confluences upstream of the Yaxchilán archaeological site facing the Lacandón Sierra Protected Area on the Guatemalan riverside, the border delineation is not as clear. The official delineation is constantly referred to, but the daily reality follows the river's rhythm. Boundaries are interpreted according to local perspectives and practices; thus, transboundarities are complex and sometimes contradictory according to riparian perspectives and perspectives based on connections or disconnections.

RESEARCHER: Well, going back to the river, how do you know where Mexico starts and where Guatemala starts?

BIOLOGICAL STATION STAFF: In this part?

RESEARCHER: Yes, in the river, in this part of the river.

BIOLOGICAL STATION STAFF: Yes, well, we know that Mexico borders Guatemala through the river …

RESEARCHER: But at what point of the river does Mexico begin?

BIOLOGICAL STATION STAFF: Well, the truth is that I don't know anything else because we see in the map of the Republic that there is a mark at the border between Mexico and Guatemala – yes, in this part. Well, we also know that because it is marked on the map that the border goes through the middle of the river, all along the middle, yes, so that is why we know that there is the other side, crossing it. Passing more than the middle of the river, we know that already, it is Guatemala …

RESEARCHER: But right now, the river is small. When it gets bigger, the middle of it is somewhere else, so it is not always the same, right?

BIOLOGICAL STATION STAFF: Well, once the river grows, we know that there are lower parts, and the river will be a little wider. Then, there are deeper parts, so logically no, it will not get wider, but it will always be the same on both sides, yes.

RESEARCHER: And when you go along the river, do you know if you are in Mexico or Guatemala?

BIOLOGICAL STATION STAFF: Well, we know, but there is no way to do it [to transit along the river without crossing the fluvial border] when the river is deeper here. It doesn't matter that we are crossing to the other side as they also come from Guatemala. It happens. If the river is deeper on this side, they must pass through Mexico – no, no, no, there is no problem. Because when the river is very low, there is a part where only the boat can pass … although you can see it, it is still wide, but it is low, and you cannot navigate.

Fieldwork, January 20, 2018

Nevertheless, conflicts arise mainly during the dry season, when the river flow decreases, sedimentary deposits emerge, and islands appear (Images 3.4 and 3.5). As explained by a Mexican riverside boatman, authorities in charge of the borders are never present, and local authorities are not involved in transboundary issues. As expressed during fieldwork by the population living on the riverbanks, transboundarities are mainly related to relations from below – in other words, daily interactions related to the river conditions without the presence of states. This presents a key difference to the international literature on the topic, such as the study of Keskinen et al. (2016), who have focused on interstate relations.

RESEARCHER: And does the government come around here?

BOATMAN: No, around here, it has never come.

Fieldwork, January 20, 2018

Sand, gravel, woods, wildlife, and natural resources are at the core of transboundary conflicts over natural resources at the shared Lacandon Forest Waterlands and the Laguna del Tigre-Gulf Riverlands. The fluvial border is negotiated and constantly redefined according to the changes of the river flows, which favors a reality of local sharing, whereas the existence of other resources provokes transboundary disputes. These resources refer precisely to the presence of sediment banks, which arise during the dry season, the transboundary smuggling of wood, and forest extractive activities.

BIOLOGICAL STATION STAFF: Well, here in the community with Guatemala, there are no problems, and people from the community can go there to the other side or without any problem. They can go walking or traveling – there is no problem. They can also cross the line and no, there are no problems. We see ourselves as friends, as neighbors.

RESEARCHER: And in the station?

BIOLOGICAL STATION STAFF: Normally, the Guatemalans here in this part of Chan Kin [a protected area] come in here a lot to cut down these trees. Recently, they took out 40 tons of wood, and they normally come here to fish here on this side of Mexico.

RESEARCHER: Doesn't it affect you – let's say – that they come into to fish?

BIOLOGICAL STATION STAFF: It doesn't affect us, but it is prohibited for them to enter here on the Mexican side because we can't cross there on the Guatemalan side [the Lacandon Sierra Protected Area]. So, there have been a few problems in that part because they enter to cut down Palma Xate and also another leaf. That is what has happened, and they have been caught, but nevertheless, they [the Mexican authorities] don't do anything.

RESEARCHER: Apart from the fact that Chan Kin is a protected area …

BIOLOGICAL STATION STAFF: Just as it is a reserve, there is also another reserve where CONANP [the Mexican National Commission of Protected Areas] also has a presence, but even so, they still do not do anything.

Fieldwork, January 20, 2018

At the crossroads of the Lacandon Forest Waterlands and the Laguna del Tigre-Gulf Riverlands, the Mexican riverside is more densely populated than that in Guatemala and more accessible by land. The alluvial plain is also the most deforested area on the Mexican riverside, although upstream of the rapids, some places on both riverbanks of the international stretch of the Usumacinta River are only accessible by the fluvial route. Consequently, the rivers are entangled with land, with the forest, and, irrevocably, with the international border. Transboundarities arise in association with the political borders but also with the locally defined boundaries, and emerge at the pace of the connections and according to the necessities of the riparian population.

What Happens to the Border When the River Disappears? The Hondo River Converted into Wetlands

The border between Belize and Mexico has never been a problem, and it may be considered one of the most politically quiet borders of Central America, mainly because the political and economic asymmetries between both states are completely accepted by Belize as part of its condition as a small state at the regional and international scales (Kauffer, 2024). This is Mexico's most forgotten border. It extends for about 190 km, 86% of which correspond to the Hondo watercourse as a fluvial border (Kauffer Michel, 2013).

The Hondo River originates in Guatemala as the Arroyo Azul, whose name clearly suggests a small and sometimes intermittent stream. The source of the Arroyo Azul (Blue Creek in English) is located in the north of Petén close to the trifinium point that marks the confluence of the three borders between Belize, Guatemala, and Mexico. The main river flows to the north, crosses the northern border of Guatemala, and then follows its course through the state of Campeche, Mexico, before crossing the land border between Mexico and Belize, where it forms the beginning of the fluvial border between both countries and is called the Hondo River. The source of the Arroyo Azul, as it is called in Mexico, known as Río Azul in Guatemala, is deep in the forest of the Maya Biosphere Reserve; the river then flows through the Mirador-Río Azul National Park, which was created in 1990, bordering the Calakmul Biosphere Reserve in Mexico and the Río Bravo Conservation and Management Area in Belize. This watercourse is also a tributary of the Hondo River in Belize.

The Río Azul in Guatemala is surrounded by lowland tropical forest in an area without human settlements except for some archaeologists' camps, as the closest village is located 50 km to the south. This national park is considered one of the best conserved areas of Guatemala, although clandestine hunting as well as fishing are facilitated by its location close to the border, with hunters and fishers entering mainly from Mexico and Belize (Moreira Ramírez et al., 2015). Nevertheless, compared with other parts of the Maya Biosphere Reserve, which have seen fires and deforestation, the isolation of and difficult access to the national park have historically protected its biodiversity.

A millennium ago, the upstream Guatemalan region of Río Azul, which today is a particularly remote location and wildlife paradise, was a prosperous lowland with an estimated population of between 200,000 and 400,000 people, including a city called Río Azul with 7,500 inhabitants. While the soils were poor and there was a lack of strategic natural resources, the location on a river discharging into the Caribbean Sea suggests a place trading with the Caribbean (Adams, 1931 /1999, 117) and "a subordinate administrative center and a fortified guardian of the frontier" (Adams, 1931/1999, xiv). Río Azul was a regional market town and an administrative center, as well as a defensive structure of the Tikal state. For that reason, the course of the river was altered and dikes and dams were built to enable navigation, and wetland gardens set up to feed a numerous population.

Located at the heart of the Caribbean Karst Waterlands, the Río Azul represents a transition zone between the dry forest of the Yucatán Peninsula and the tropical forests of the Lowland Maya Forest (García & Radachowsky, 2004). The specific waterlands of the Río Azul feature two types of aquatic habitats. The river consists of "a series of disconnected sections of stagnant water, from 100 meters to a few kilometers long, and approximately 10 to 30 meters wide" (García & Radachowsky, 2004, p. 7). Along the "river course", there are ponds or water holes, known as *aguadas* in Spanish, which are small, stagnant pools several meters deep and not exceeding 30 meters in width.

In the past and today, the dry season at the upstream Río Azul in Guatemala causes wetlands, creeks, and ponds, also known as *pozas* in Spanish, to dry out, especially between February and May (Baur, 2004). Consequently, the Río Azul stops flowing during the dry season and becomes a set of ponds or pools (Baldizón & Bravo, 2004). This affects not only the upper sections of the Arroyo/Río Azul but, also the part of the border where the Hondo River delineates the political boundary between Belize and Mexico.

Owing to the lack of a treaty between Belize and Mexico, the land border between the two states is not clearly demarcated and has no physical border markings. In 2005 and 2006, the Mexican government proposed a treaty that also included the maritime border, without obtaining a positive response from Belize. Nevertheless, the absence of demarcations is not a political and geopolitical problem. However, a 2007 audit by the Mexican federal government based on a fieldwork survey found that the so-called meridian of Garbutt (i.e., the terrestrial line between both states) was not correctly located on the maps (Auditoría Superior de la Federación, 2007). Nineteenth-century maps and today's geographic information systems do not agree with each other but, owing to the lack of settlements, this is not a substantial problem.

When the Hondo River first becomes an international watercourse with the border defined as the thalweg, the stream is small and presents the same characteristics as the upstream sections of the Río Azul: during the dry season, the river stops flowing and tends to turn into wetland. Consequently, the thalweg disappears, and the border evaporates. The disappearance of a river border could be considered ironic insofar as the existence of a border legally defines the territory of a state. In our case study, however, beyond the need for both governments to work on a future treaty to define an adequate common border respecting the local conditions, it enables us to analyze the effects of this situation within the framework of the concept of waterlands and its relationships with transboundarities.

The Hondo River is the major surface watercourse of the Yucatán peninsula, holding 97.4% of its subterranean waters and characterized by a karst formation that combines a lack of surface deposits with a subterranean system of thousands of *cenotes*, *poljes*, and caves, ranging in depth from hundreds of meters to tens of kilometers. Caves and *cenotes* play a fundamental role in sustaining the ecological niches of the region as they support the local ecosystems.

The Hondo River is deeply meaningful from our Maya Forest waterlands perspective. On the Río Azul (i.e., the upstream section of the river), there is clear evidence of the interactions between the presence of archaeological sites and today's conservation initiatives. The human activities in this territory are focused on two aspects: extraction and conservation of natural resources (hunting and fishing versus conservation initiatives) and the interest in discovering and understanding ancient civilizations through archaeological excavations. In both cases, the boundaries between lands, forests, and waters are fluid and constantly changing, according to the climate but also to the presence of humans, which favors entanglements.

The upper part of the Hondo River in Guatemala belongs to a climatic and geologic transition area between the dry Yucatán peninsula, which has almost no surface waters, and the Caribbean Sea. It also forms an ancient bridge, as well as a frontier area and human shield. The upstream Hondo River represents a past site of exchanges and human circulation as well as a boundary that has now been converted into a frontier area located in the vicinity of the confluence of the borders of the three states that belong to the Maya Forest Waterlands. Regarding transboundarities, it presents a highly complex situation of entanglements owing to the multiple fluidities that connect climate, geology, waters, and forests, as well as past and present.

Going downstream on the Hondo River, the historical characteristics of these entanglements between waters, lands, and forests that define this specific and core site of the Caribbean Karst Waterlands are also entangled with the current political border; however, this does not represent a conflict, although it is an unresolved border issue according to the Mexican government. At this point, transboundarities associated with a clearer political boundary denote the national characteristics of local populations and the different perspectives on the "others" living beyond the border.

When Wetlands and Swamplands Are Entangled with the Border: Disputed Waterlands at the Sarstoon River

As previously explained, the political boundary between Guatemala and Belize will be referred to as the "adjacency line" when dealing with Guatemala and as "the border" from the Belizean perspective. When we convey a general perspective, we choose to use the notion of a political boundary. This issue is omnipresent when we mention the relations between both states, but also between nationals from both countries. On several occasions during Kauffer's fieldwork in Belize, our Mexican team was mistaken for Guatemalan nationals and clearly considered unwelcomed by Belizeans.

In this complex sociocultural context, although without normative existence, the issue of the border between Belize and Guatemala is politically edgy and ubiquitous.

The political boundary comprises two sections: the land boundary and the lines drawn in the international part of the Sarstoon River. The land boundary consists of a line called, in Mexico, the "Garbutt's Falls Meridian" and is mentioned as such in the Mariscal Treaty between Mexico and British-

Honduras (SRE, 1893) and referred to in article 1 of the Aycinena-Wyke Treaty between the Republic of Guatemala and Great Britain, signed in 1859 (Akerman, 2017), which supposedly delineated the boundary. It is defined as

> beginning at the mouth of the River Sarstoon in the Bay of Honduras and proceeding up the mid-channel thereof to Gracias a Dios Falls; then turning to the right and continuing by a line drawn direct from Gracias a Dios Falls to Garbutt's Falls on the River Belize and from Garbutt's Falls due north until it strikes the Mexican frontier.
>
> Akerman, 2017

The Garbutt's Falls Meridian is a straight line that extends from the south to the north and divides Belize on the east from Guatemala on the west. When the line meets the Sarstoon River, the border, from that point on, coincides with the international river Sarstoon.

The Sarstoon River originates in Guatemala in the Alta Verapaz Department in the Santa Cruz Sierra as the Chahal River and changes its name when merging with the Chiyú River. It is then called Gracias a Dios River until the next confluence with the Franco River, where the so-called Sarstoon River and the border begin. The river flows along 111 km until it reaches the Caribbean Sea – 55 km of which coincide with the border. It widens downstream, and close to its mouth is Sarstoon Island, which represents a major problem in the contemporary disputes over the precise location of the boundary. As stipulated by the Treaty of 1859, the political boundary on the river is located on the mid-channel. Nevertheless, while the dispute is ongoing, the two countries do not agree on the interpretation of its exact location. In the case of Sarstoon Island, Belize considers itself to have sovereignty over the island, which is contested by Guatemala. The Treaty of 1859 has not been recognized by Guatemala because certain terms were not fulfilled neither by Britain nor by Belize – which promised to build a road to the Caribbean Sea – and this non-compliance forms part of the legal basis of the Guatemalan territorial claims. Additionally, different sources that mention the diverse skirmishes present different interpretations of the boundary delineation: some argue that it is on the middle line, and others consider that it corresponds to the thalweg.

Consequently, border incidents at the Sarstoon River are a common occurrence. The Guatemalan Army is on constant patrol with weapons, preventing Belizeans from advancing along the river, while the Belizean Defense Forces try to move away from the area to avoid skirmishes. However, a group of citizens known as the Belize Territorial Volunteers (BVT) have taken it upon themselves to defend the territory and oppose the Guatemalan forces. In February 2024, a journalist from the *New York Times* witnessed an incident between civilians from Belize and the Guatemalan Army, which involved rifles (Romero & Cegarra, 2024).

Among the latest incidents, on April 15, 2019, members of the Coast Guard of Belize were intercepted by the Guatemalan Navy, which prevented them from advancing further on the river (Breaking Belize News, 2019). On July 23,

2019, the Guatemalan Armed Forces stopped a Belizean boat from accessing the river, which caused an incident with the Belize Defense Forces (Belize Press Office, 2019). On September 18, 2022, five maritime vessels of the Guatemalan Armed Forces were found on the northern part of Sarstoon Island. The Belizean flags put up by the BVT were removed by the Guatemalan Armed Forces (The San Pedro Sun, 2022). In October 2023, the Prime Minister of Belize appealed to the public asking to avoid more conflict with the Guatemalan Army over the Sarstoon River, as the state was not able to give security guarantees, which was a clear reference to Belize's powerlessness when faced with threats by Guatemala, resulting from its condition as a small state (Caribbean News, 2023). This list could go on.

The Sarstoon Temash National Park (STNP) was created by the government of Belize in 1994 and was also recognized as a Ramsar site in 2005. The transcript (1998) of a stakeholders' workshop, held in 1997, indicates that the area was considered as having limited farming potential, but the finest mangrove forest in Belize. The nearby communities had an interest in forming a reserve in 1989, when there was a suggestion for a biosphere reserve centered around Punta Gorda, Sapodilla Cayes, the Columbia Forest, and the Temash River. However, owing to a lack of funding, the idea did not succeed, but instead, a national park was created in 1994. In 1997, a stakeholders' workshop was organized with the Indigenous and Garifuna communities living around the park. The latter argued that they could take part in co-managing the area as equal partners, as they were part of the ecosystem which generated the rich biodiversity of the wetlands.

Since 1997, the national park has been formally co-managed by the Sarstoon Temash Institute for Indigenous Management (SATIIM), an Indigenous organization, which has had two specific tasks: to co-manage the park and to defend Indigenous land rights. The Protected Area Management Plan published by SATIIM (Herrera, 2004) emphasizes the border issues related to clandestine hunting, highly destructive fishing, and the illegal harvesting of forest products by people crossing the border, as well as the obstacles to building a corridor of connectivity due to the deforestation in Guatemala. The plan mentions the reality of encroachments by Guatemalan nationals and the difficulty in resolving this issue owing to the border dispute – a situation also deplored by the local Belizean population. The presence of the Guatemalan Army, seen as a threat for Belize, does not encourage effective patrolling to dissuade "borders violators", as SATIIM (2004, p. 74) called them.

Fieldwork and interviews with different conservationists in Belize between 2022 and 2023 indicated that, since the 2015 Consent Order issued by the Caribbean Court of Justice concerning the Indigenous customary land rights in Belize, SATIIM had withdrawn from the national park co-management to focus on the process of demarcating the Mayan lands. As described in Chapter 2 regarding the Columbia River Forest Reserve, in a similar vein, the Consent Order affects – and leaves open – the management of the national park as its territory coincides with the Indigenous customary lands. Thus, Indigenous organizations and communities have withdrawn from protected area co-management as they consider that the lands corresponding to the protected area are subject to customary rights and therefore pertain to them. This means that the protected areas now exist in a vacuum, as the government has not implemented the Consent Order. In the case of the Sarstoon-Temash National Park, the situation might also bring about the disappearance of the Ramsar site, as it is based on its classification as a national park.

The same fieldwork and interviews with various conservationists also indicated that the area continues to face growing pressure for natural resources and some expanding milpa cultivation. On the Belizean side, the interviewees mentioned that there was no site management on the Guatemalan side either, and that the co-managing NGOs had withdrawn because of threats. Nevertheless, a representative of the Belize Biodiversity Office stated that SATIIM had detected illegal activities in the area, although the scope of these was unknown. The representative mentioned the lack of institutional presence in this border area with Belizean vessels being constantly stopped and blocked by the Guatemalan Army. As a result, this border area was considered the riskiest and most complicated, despite the existence of diplomatic relations. A conservation organization in Punta Gorda commented that they used to go out on patrols together with the co-managing partner Foundation for Ecodevelopment and Conservation (FUNDAECO) on the Guatemalan side; however, they stopped these activities and are now focused on research exchanges.

Downstream on both riversides the river feeds an important wetlands area that was converted into a protected area and into Ramsar site (i.e., area from the List of Wetlands of International Importance derived from the Ramsar Convention of 1971). The description of Ramsar site states that it offers

> a complex of several different terrestrial ecosystem types located on the southern frontier with Guatemala, bisected by two large rivers, one of which forms the border. Seasonally and permanently flooded forests predominate, with some 1,100 hectares of lowland sphagnum moss bog unique to the region, a saline/brackish inland lagoon, and 9,600 ha of saline swamps, with the country's most undisturbed and largest stand of red mangrove (*Rhizophora mangle*) and its only stands of Comfra Palm (*Manicaria saccifera*).
>
> Ramsar, 2024b

On the opposite riverbank is the Sarstoon River Multiple-Use Reserve (Reserva de Usos Múltiples Río Sarstún [AUMRS], in Spanish), which was declared in 2005 by Guatemala and accepted as a Ramsar site in 2007. The description highlights that it is

> located along the southern border with Belize and adjacent to the Amatique Bay (Image 3.7). The reserve is formed by a series of wetlands, ranging from continental and coastal to artificial. It has a transboundary character, since it acts as a buffer zone for the wetland of the Sarstoon-Temash

Image 3.7 A View from Belize Towards the Trinational Amatique Bay
Source: Hanna Laako, 2023

> Ramsar site in Belize. It is an important stop-over and breeding site for migratory waterbirds, including several flagship species. It also assists in the regulation of the local microclimate and promotes other hydrological processes, including aquifer recharge. It possesses the remains of the Caribbean Biological Corridor ecosystems and karstic wetlands that have unique characteristics.
>
> Ramsar, 2024a

The reserve is administered by the National Council of Protected Areas (*Consejo Nacional de Áreas Protegidas* [CONAP]) – a national government institution in partnership with a Guatemalan NGO, FUNDAECO, which is part of a consortium with the Indigenous peasant association "Amantes de la Tierra". This consortium was set up following land conflicts between the local population and the Guatemalan government.

We find the contrast between the Belizean perspective and the Guatemalan one interesting. Guatemalans, who always, during fieldwork, express major concerns about the border and constantly demand for sovereignty to be respected, refer to the Ramsar site as transboundary (Ramsar, 2024a) as well as of "binational importance". Consequently, in information provided by Guatemala regarding this conservation area, transboundarity is presented as an opportunity. Both states have declared a protected area and have registered a Ramsar site, but each one administrates its territory independently, although in recent years, the idea of transboundary cooperation has progressed (see the next part of this chapter). For Belize's stakeholders, the border situation at the Sarstoon is the main issue regarding the threat to conservation and illegal activities linked to clandestine incursions from Guatemala.

The Sarstoon River shapes an estuary system that includes mangroves, rivers, lagoons, and flooded forests and is usually flooded by both the tide from the Caribbean Sea and continental overflows (FUNDAECO, 2024), representing a clear example of coastal waterlands characteristic of the Karst Caribbean Waterlands corridor. The local population is made up of ten Q'eqchi' communities on the Belizean riverside and 30 on the Guatemalan riverside, with most of them Q'eqchi's, but also some Mestizos, Garifunas and Chortis. In Barra Sarstún south of the Sarstoon island on the Guatemalan riverbank, where the river merges with sea waters, fishing is the traditional activity; however, in 2013, local fishermen declared Barra Sarstún a recovery area owing to the diminishing fish stocks and diversity. From that moment, various groups have participated in mangrove protection and restoration projects financed by NGOs, international and national projects, as well as national subsidies awarded to local organizations such as the Mayan association for Wellbeing in the Sarstoon Region (Asociación Maya pro Bienestar Rural del Área Sarstún, APROSARSTÚN).

As can be seen, the Sarstoon River as a border is embedded in multiple "transboundarities", meaning overlapping and vague dynamics related to border disputes and those between Indigenous people and the Belizean government over customary lands and protected areas. In this case, in waterlands where forests, the river, and associated ecosystems are part of strategic interests that define both connections and disconnections, conflicting transboundarities are expressed as land claims and disputes between Indigenous organizations and states.

Transboundary Conservation in the Maya Forest: Political Borders, Ecosystemic Borderlands, and Hidden Waters

In Chapter 1, we extensively discussed the emergence of eco- and bioregional mapping by conservationists, ecologists, and biologists, which was later accompanied by the designation of biodiversity hotspots to identify priority conservation areas. According to Zbicz (2003), the ecosystem level of conservation management was recognized as the most effective way to protect habitats and biodiversity. Additionally, this led to the need to transcend both political and protected area borders. Thus, a strategy of transboundary conservation emerged and has been actively promoted since the 1990s.

Transboundary conservation as such is not a new phenomenon. According to Mittermeier et al. (2005), the USA and Canada created a transboundary protected area around the Waterton Glacier in 1932. Thereafter, the concept of transboundary conservation was introduced in the 1974 First Central American

Meeting on the Management of Natural Resources and Cultural Resources, which emphasized the benefits of joint management in border areas between countries. Between 1979 and 1981, Columbia and Panama introduced protected areas along their border, which were later designated as UNESCO heritage sites. In the 1990s, a global strategy was developed to address *environmental peacebuilding* in border conflict areas. Simultaneously, the number of so-called transboundary conservation areas grew significantly (Mittermeier, 2005). On the one hand, it was acknowledged that environmental legislation was mostly designed for peaceful times; on the other, it was argued that designating areas with border conflicts as peace parks would promote peace (e.g., Ali, 2007). This has led to lively debates in social sciences about the war–peace–conflict axis of the strategy (see, for further discussion, Laako et al., 2022).

However, transboundary conservation is not only troubled by war and peace. One of the most ambiguous aspects of the strategy is the definition of transboundarity. As pointed out by Brenner and Davis (2012), transboundary protected areas are rarely transboundary (i.e., they would formally span a jurisdictional border). In fact, transboundary protected areas are predominantly internationally adjoining protected areas (IAPAs). Transboundary protected areas have referred to "protected areas contiguous across a national boundary, clusters of protected areas with intervening land, clusters of separate protected areas without intervening land, transborder areas that include proposed protected areas, and protected areas in one country with minimal or sympathetic land use in the adjacent country" (Brenner and Davis, 2012, p. 501). According to Mittermeier et al. (2005), transboundary conservation refers to an area that straddles international boundaries and is managed jointly for conservation purposes. For Erg et al. (2015), transboundary conservation can include multiple types of collaboration, which may or may not be strictly intergovernmental. In our previous work (Laako et al., 2022), we divided transboundary conservation into two categories: the first category we called "transboundary protected areas", on the basis that these areas had been recognized by UNESCO as transboundary; in other words, the reserves were subject to some sort of bi- or multilateral agreement. Most of these are UNESCO heritage sites, biosphere reserves, or Ramsar sites and located in Europe. The second category we called "complexes of transboundary conservation", which were based on IAPAs and mostly found in tropical forest areas and within biodiversity hotspots, such as the Maya Forest. We concluded that while important transboundary collaboration and promotion of biocultural landscapes took place in these areas, they mostly served to strengthen international borders rather than superseding them.

This also indicates the problem of protected areas as potentially isolated enclosures. As Zbicz (2003) and Brenner and Davis (2012) have highlighted, the focus of transboundary conservation should be on ecological and human cross-border interactions, rather than on the borderland location of the reserves. In this chapter, we consider another angle related to the concept of transboundarity: rethinking the different ecological boundaries that are crossed by organisms as driving forces. Talley et al. (2006) have addressed interhabitat connectivity between land and water interfaces as a relatively new area of ecological research, which explores "transboundary" links between distinct habitats, such as forests and grasslands, freshwater and marine habitats, or aquatic and terrestrial ecosystems, thus creating their own "eco-borderlands". Anthropogenic factors may also influence the connectivity of organisms crossing these ecological boundaries, thus altering them.

Talley et al. (2006) have focused on the eco-borderlands or transboundarity between terrestrial and aquatic ecosystems (i.e. forest waterlands of sorts). They have argued that the identification and rethinking of water–land interfaces is increasingly important for conservation, especially because these "transboundary spaces" have been affected by anthropogenic activity – for example, the loss of boundary areas, such as marshlands, between land and sea. Another example is the channelization of river basins and transformation of lakes into reservoirs, which has influenced the connectivity between land and water by restricting natural flows and preventing flooding, thus eroding wetlands ecosystems. Roads and highways have also created boundaries for species connectivity between ecosystems. Land–water interfaces, especially coastal zones, but also riparian areas, lakes, and floodplains, are the most biologically diverse areas on earth. Yet, the authors have highlighted that connectivity is not automatically a virtue, because it may also cause different kinds of problems to conservation, including habitat destruction.

While interhabitat connectivity has been recognized since the 1920s, research in this area has markedly increased since the 1980s, with the introduction of landscape ecology. However, "transboundary research" into these interfaces continues to be challenging, as studies and funding tend to be divided as to whether they deal with research on terrestrial ecosystem or aquatic ecosystem units, in a similar vein as the social sciences tend to be bound to national borders.

This part of the chapter addresses transboundarities in terms of the Maya Forest conservation with a particular emphasis on waters. It first shows how transboundary conservation extends toward the connectivity of ecosystems and species, although the issue of waters remains hidden. As the Maya Forest is a concept constructed by ecologists and conservationists on the one hand and archaeologists and anthropologists on the other, there has been an emphasis on forests and Mayan ruins, as elucidated in Chapter 2. As mentioned in the introduction, the strategy of transboundary conservation at the global level has been focused on forests, because many of the remaining, biodiversity-rich forests are located in international borderlands (Laako et al., 2022). Indeed, our previous research results have suggested an important correlation between tropical forests, biodiversity hotspots, and transboundary conservation complexes formed of IAPAs. For these reasons, in the following, we explore transboundary conservation in the Maya Forest in terms of water dimensions. The latter part of the section addresses the transboundarities and (dis)connections between the forest–water nexus and ecosystems, showing how transboundary conservation itself is challenged in these respects.

Maya Forest Transboundary Conservation: Beyond Political Borders, Toward Hidden Waters

At the macro-level, there have been two major transboundary conservation projects in the Maya Forest: the Mesoamerican Biological Corridor MBC (e.g., Carrillo et al., 2022; Finley-Brook, 2007) and the Selva Maya megaproject (e.g., Laako et al., 2022). According to interviews carried out between 2022 and 2023, the latter project builds on the former. The MBC existed for 18 years (2000–2018) and involved intergovernmental collaboration in Mexico and Central America, as well as building and fostering conservation NGOs, their networks, and local participation (e.g., Laako & Kauffer, 2021). Especially in Mexico, the MBC's interpretation of connectivity was changed from an exclusive focus on creating more protected areas to a view of landscape connectivity involving collaboration with communities.

A version of this emerged with the Selva Maya megaproject funded by the German Agency for International Cooperation GIZ. In fact, for many current conservationists, the concept of "Selva Maya" (Maya Forest in Spanish) is exclusively related to this particular megaproject. The Selva Maya project initiated at the beginning of the 2000s has centered on the trinational "heart" of the Maya Forest: the Maya Biosphere Reserve in Petén, Northern Belize, and Campeche and Quintana Roo in Mexico. The interviews with conservationists and governmental agencies between 2019 and 2023 indicated that the project is predominantly focused on creating connectivity between protected areas via transboundary collaboration in the form of cross-border ranger meetings, intergovernmental and inter-sector reunions and agreements, and some shared patrolling, albeit with restrictions, given the border disputes between Guatemala and Belize. For a long time, conservationists have pointed out that formal transboundarity between the two countries is not possible, as such an endeavor would mean acknowledging the border, which is in dispute. This shows the challenge of transboundary conservation aimed at peacebuilding: when countries do not agree on a border, jurisdictional or political transboundarity is not an option from their viewpoint. As long as the borders are in a vacuum, no transboundarity is possible.

Thus, the Selva Maya megaproject is more about ecosystem and species connectivity, whereby each actor is mainly focused on their side of the border, hoping that the ecosystems expand beyond political borders and that species move across and along them.

In this sense, the recent development of the Maya Forest on the Belizean side is an important component of the Selva Maya project. The Maya Forest Trust Fund, a consortium of a dozen conservation organizations, was formed in 2021, when 96,000 hectares of private land located next to the Río Bravo Conservation and Management Area were purchased. Additionally, the Maya Forest Corridor (previously called Central Belize Corridor) has been purchasing smaller areas of land in Central Belize to enable species connectivity toward Northern Belize and between the northern and southern parts of Belize.

According to a representative of the Belizean Biodiversity Office interviewed in 2022, this corridor-thinking builds on the MBC and Selva Maya project and involves "a set of big conservation players, such as Re:wild and Nature Conservancy" to connect with the trinational heart of the Maya Forest.

A conservation leader in Belize, interviewed in June 2023, explained that the new protected area of the Maya Forest Trust Fund, comprising privately purchased land, was the result of a collaboration built over a long time – at least the past 18 years. It was started in the context of various mounting pressures related to fires and forest clearing carried out by the Mennonite communities located in El Cayo and Orange Walk, as well as the increasing problem of smuggling from the Hondo River (Image 3.8). Additionally, the foreign privatization of lands in Belize has been a worry, and hence, when the land became available for sale, the Trust Fund was formed to purchase and protect it. This conservation leader argues that the main objectives of the Belizean Maya Forest Trust are the protection of the health of its forest and water ecosystem, the aquatic ecosystem connected to the Hondo and Belize River basins, the connectivity to enable corridors for wild cats, and the conservation of the unique ecosystem of the Cara Blanca Pools, which consist of lagunes and *cenotes* along the Yalbac escarpment. As can be seen here, water is, or has become, a key element, although hidden beneath the Maya Forest vocabulary.

Indeed, the Selva Maya megaproject has also engaged with a seemingly hidden biodiversity conservation of forest waterlands: the monitoring and

Image 3.8 Mennonite Blue Creek Dam at the Belizean-Mexican Border
Source: Hanna Laako, 2023

protecting of *aguadas*, or waterholes (Selva Maya, 2019; 2021). These have been identified as a prime indicators of biodiversity conservation and the effects of climate change, as many species of wildlife, such as peccary, tapir, puma, and jaguar, some of them endangered, depend on them. For these reasons, since 2008, the project has monitored and systematized information about waterholes and the associated fauna in the Maya Forest.

Laako's interviews with conservationists in the region between 2022 and 2023 also revealed some criticism of the Selva Maya megaproject. Many commented that most of the support was technical and lacked a more inclusive approach to transboundary conservation. Some conservationists also disliked what they perceived as an increasingly militarized, masculine, and equipment-based approach to patrolling and surveillance; some others, however, supported such an approach.

In addition to these intergovernmental megaprojects, the interconnections between transboundary conservation and the waters in the Maya Forest have appeared in the earlier proposals of peace parks and within some projects of smaller conservation organizations. As indicated above, peace parks have been proposed along the Belize–Guatemala border together with biosphere reserves, but have not yet been set up.

Again, smaller NGOs have joined forces to improve biodiversity conservation, which addresses some transboundary challenges. For example, Castro (2022) has written that connectivity between protected areas across political borders is important to sustain biodiversity and freshwater refill that the humid tropical rainforests, such as the Maya Forest, represent. As government priorities may shift and projects end, as it happened to the MBC, Castro (2022) has maintained that collaborative work between civil society organizations focused on conservation is paramount. Thus, their organization, *Natura Mexicana*, has engaged in transboundary collaboration with partners from Belize and Guatemala to allow species connectivity, especially for the scarlet macaw and the jaguar. They have, for example, formed a Conservation Alliance for Scarlet Macaws in the Maya Forest, funded by the Selva Maya megaproject. They have also become part of the jaguar-habitat conservation alliance *Jaguares de la Selva Maya*, which seeks to create habitat and corridor connectivity between Chiapas and Guatemala. They also address the potential human–wildlife conflict (e.g., De la Torre et al., 2019). Yet again, waters may be hiding in these projects, given that jaguars do not survive in large territories without waterholes.

A new megaproject called *Mesoamerica's Five Great Forests* has recently been set up under the leadership of the Wildlife Conservation Society and has received considerable funding from the European Union (WCS, 2022). The project addresses climate change in five forests, which run from the Maya Forest in Belize and Guatemala to La Moskitia between Nicaragua and Honduras, Indio Maíz-Tortuguero in Nicaragua and Costa Rica, La Amistad in Costa Rica and Panama, and El Darién in Panama and Colombia. All of these five great forests are located in borderlands between countries. While the focus here is on forests, the project also addresses climate resilience, which suggests the consideration of waters. However, as the project is new, it is too early to draw conclusions about its results.

Fieldwork and interviews between 2022 and 2023 indicated that, at the grassroots level of conservation, waters were considered increasingly important, although this was not reflected in the predominant Maya Forest vocabulary. Some conservationists have pointed out that the term Maya Forest, as many other terms, has been created to attract funding and tourism interest in ruins within the jungle. In practice, conservation work often deals with waters – in particular, the monitoring and researching of water quality, the effects of pesticides, and invasive species. Problems of water scarcity in transboundary contexts have occasionally been mentioned. The link is particularly strong in the Maya Forest coastal areas, along the Mesoamerican Reef. Down the River Hondo, in the Banco Chinchorro Biosphere Reserve and the Manatí Sanctuary, conservationists constantly have to deal with both transboundary issues and the connectivity between marine and terrestrial ecosystems in a similar vein as around the Port of Honduras Bay. In addition to the problem of government projects addressing transboundary issues being short-lived, conservationists also mention that the public is often hostile toward transboundary efforts, suggesting that these kind of attempts should be exclusively concentrated on their own country (see also Chapter 5).

Biodiversity Conservation in Ecosystemic Borderlands? Challenging the Forest Waterlands' Transboundarities in the Maya Forest

Since the 2010s, there has been an increasing tendency toward coastal, marine, ocean and wetlands conservation. However, as indicated in the introduction, these categories are still predominantly dealt with separately, as terrestrial and aquatic conservation. In terms of transboundary conservation, Mittermeier et al. (2005) created a marine area of the Maya Forest in its own category, which they called the "Mesoamerican Reef". The Mesoamerican Reef, a distinct transboundary conservation complex, was defined as the "Caribbean jewel", running along the Yucatán and Quintana Roo Coast to the coastal areas and reefs of Belize and ending at the Bay Islands of Honduras. It also comprised IAPAs, followed the shallow Caribbean waters buffered by the reef and mangrove forests. It included a protected, navigable seaway, which was considered to include the Sapodilla Cayes and parts of the Sarstoon-Temash National Park discussed above. The Belizean Mesoamerican Reef and Mexico's Sian Ka'an Biosphere Reserve were recognized as the most valuable protected areas and heritage sites by the Tulum Declaration of 1997, in which the heads of state of Mexico, Belize, Honduras, and Guatemala pledged to support "the protection and sound use of the shared coastal areas" (Bezaury et al., 2005). Yet, according to Canty et al. (2018), despite improved national legislation, seven ratified international and regional agreements, and 43 protected mangrove areas in the transboundary Mesoamerican Reef, the mangrove forests, which can be defined as certain types of forest waterlands and ecosystemic borderlands, are eroding.

The main problems are, according to Canty et al. (2018), a lack of governance framework transparency, the disconnect between research and management, and geopolitical differences.

Hence, the Mesoamerican Reef has been identified as a separate unit from the Maya Forest, with two sets of transboundary conservation strategies that illustrate the disciplinary difficulty of crossing conservation boundaries between terrestrial and aquatic habitats, despite the conservationists' insistence on overcoming borders for biodiversity connectivity. In the case of the Maya Forest, the separation results in the waters "being hidden" to a certain extent and a disconnection of the forest waterlands. During fieldwork in Belize, this was well illustrated in an encounter with a group of North American Ecology students and professors in a biological station located in Chiquibul National Park. The previous week, the group had been staying in another biological station located on the Mesoamerican Reef. The professor explained to us that such a combination of field trips was exceptionally rare and that nowhere else than in Belize there was marine and reef biodiversity in such close proximity to humid, tall tropical rainforest. In his view, this was a remarkable opportunity for the students to explore these two separate ecosystems, which were located so close to one another they could be visited in the same field trip. However, it remains unclear to which extent the connectivity between these two habitats was explored, not to mention the eco-borderlands that remain between the two ecosystems, formed of marshlands and the more heavily built-up environment of Belize (i.e., its ecosystemic borderlands and forest waterlands between the specific marine and rainforest habitats). There are now plans to include these areas in the Maya Forest Corridor, which aims at enabling jaguar movement, among others.

The difficulties of rethinking transboundarities are also illustrated in the academic literature on the Maya Forest, as has been mentioned also in the introduction. Although attempts have been made to bring together a broader and less border-bound analysis, which is the very foundation of the Maya Forest concept, the literature often consists of compilations, to which each scholar or conservationist contributes a separate piece, examining a subject either from one side of a political border or an ecosystemic border (e.g., Primack et al., 1998; Gómez-Pompa et al., 2003) This is perfectly understandable. It takes time and resources to study the depths of one's own field and context to achieve meaningful results, without risking to digress toward broad generalizations that often remain superficial. Crossing political borders into other cultures, languages, and dynamics while resolving the logistics of permits is something that many scholars avoid, as pointed out by Primack et al. (1998). Sticking to one's own well-demarcated terrain has many benefits; this is particularly true for the Maya Forest, where scholars and conservationists have actively sought to engage with one another, but ended up eventually just working on their side of the border, because doing otherwise has proved to be too difficult. As an academic commented in an interview in Quintana Roo in June 2023, cross-border research and regional studies have often been undertaken by North American scholars because they can afford them. Moreover, there are also academic challenges within the countries of the Maya Forest as some academic institutions in Mexico are more influential than those in Petén in Guatemala, or in Belize. This also means that research and knowledge on the Maya Forest is easily taken outside the region, which is also an argument raised by many interviewees during Laako's research in the region. The rethinking of transboundarities needs to consider global academic power relations, as the Maya Forest is also a scholarly endeavor and the challenges related to transboundarities are not only political or ecological, but also academic.

Despite the obvious challenges of rethinking transboundarities connected to our political, social, or disciplinary borders, substantial attempts have been made to enter the Maya Forest waterlands. Our research on the transboundary conservation of the Maya Forest confirms that the protected areas do not span jurisdictional borders but are formed of IAPAs, with each one being predominantly focused on one side of the border. However, the protected areas themselves as a category tend to extend beyond ecosystem divisions. In other words, protected areas include forests and waters – categories often otherwise separated in national legislations. Additionally, given their tendency to focus on connectivity and corridors, conservationists have actively looked beyond protected area borders to take account of landscape continuity and movement.

Conclusions

There are neither treaties nor international laws regarding transboundary cooperation on water issues between states for transboundary and international rivers, and wetlands at the three above-mentioned borders between Guatemala–Mexico, Belize–Mexico, and Belize–Guatemala. Along the political borders of the Maya Forest waterlands, the border, its history, and its entanglements with the rivers – and in some cases with the sea – delineate specific borderlands. These borderlands share common dynamics, but they also shape the different types of Maya Forest waterlands. Nevertheless, as we have shown, local situations may differ on each riverside and in each village depending on the forests, waters, people, any specific boundary and transboundary issues, and the impacts of the effects of interstate relations. Consequently, transboundarities are complex, heterogeneous, and shaped by water and border entanglements, and they are mainly connected with the forests.

Transboundarities in the Maya Forest waterlands do not specifically focus on interstate relations as part of the international literature tries to suggest (Keskinen et al., 2016), but they are entirely defined by the international border and its entanglements with waters, forests, and past and contemporary international relations. They are, above all, based on daily local interactions and on the presence of stakeholders interested in water and conservation issues, sometimes focusing on the former and sometimes on the latter. Consequently, transboundarities are expressed in terms of connections, *misconnections*, or disconnections – but above all, as connectivities.

This chapter highlighted three main domains of transboundarities: the nature-based ones linked with the fluidity of waters and the connectivity of biodiversity, which transcend human boundaries (i.e., political or social ones); the fluctuating local transboundarities, which depend on the riparian population and on the result of their relations with waters and forests, and of their entanglements with the political border; and the misconnections or conflicting transboundarities, which arise from the contradictions of the boundary history, the existence of old and new claims over lands, waters, and the states' sovereignty, and connect or (sometimes) disconnect.

Given the porous boundaries of waterlands corridors defined by our research and the existence of lands claims, conflicts about waters, and interests in forests, transboundarities are based *in situ* and constantly transforming. This chapter portrayed some leads among the Maya Forest waterlands from recent fieldwork.

The topic of rethinking transboundarities and connectivity – especially in the case of the Maya Forest waterlands – is not limited to the examples shown in this chapter, but instead, it sheds light on potential further research. One important issue for further research, which could only be mentioned in this chapter, is the nexus of forests and waters; this is a fairly recent but emerging topic. Thus, the literature on tropical rivers in Geomorphology (Latrubesse et al., 2005; Syvitski et al., 2014), and Ecology opens further perspectives on the water–land nexus through habitat connectivity and the analysis of aquatic and terrestrial ecosystem linkages (Talley et al., 2006).

However, our discussion of the Maya Forest transboundarities shows an emerging consideration of the forest–water interface, particularly in conservation. This is illustrated, for example, by the projects related to monitoring waterholes for jaguars and the growing interest in examining "ecosystemic borderlands", which involve many different forest waterland connections and biodiversity.

Another future topic beyond the scope of this chapter is ecotourism as a substantial human activity related to borders and the many kinds of transboundarities in the Maya Forest. Our previous research and fieldwork on the Mexican side of the Maya Forest have found an increasing tendency of conservationists and locals to develop ecotourism, defined as "responsible travel to natural areas that conserves the environment, sustains the well-being of the local people, and involves interpretation and education" (TIES, 2015). This not only allows raising funds for conservation and local communities combined with promoting biocultural awareness, but it also enables different forest waterlands connectivities. While the attraction of the Maya Forest is based on combining Mayan ruins and forests, many conservationists commented in our previous research that, as tourists would not come purely for archaeological sites without vegetation, the same applies to water. Tourists are interested in access to water to enjoy the landscape – rivers, *cenotes*, or waterfalls – or in leisure activities involving water such as canyoning or canoeing. Our fieldwork suggests that most ecotourist locations, such as those in the Lacandon Forest Waterlands, are situated in the vicinity of rivers, although some are also in a state of abandonment owing to the challenges to maintain the activity and attract backpackers. While the discussion on ecotourism as an activity to enhance forest–waterland interfaces and conservation could not be elaborated further in this chapter, we have identified this topic as a unifying nexus that warrants further exploration.

References

Adams, R.E. (1931/1999). *Río Azul: an ancient Maya city*. University of Oklahoma Press.

Akerman, J.R. (Ed, 2017). *Decolonizing the map: Cartography from colony to nation*. University of Chicago Press.

Ali, S. (Ed, 2007). *Peace parks: Conservation and conflict resolution*. MIT Press.

Auditoría Superior de la Federación (ASF). (2007). *Informe del resultado de la revisión y fiscalización superior de la cuenta pública 2007*. Tomo II volumen 2, Auditoría Superior de la Federación.

Baldizón & Bravo (2004). Fish. In García, R. & Radachowsky, J. (Eds). *Ecological evaluation of the Mirador-Río Azul National Park, Petén, Guatemala*. Wildlife Conservation Society. https://www.researchgate.net/publication/284730660_Ecological_Evaluation_of_Mirador-Rio_Azul_National_Park_Peten_Guatemala.

Baur, E. (2004). Large mammals. In García, R. & Radachowsky, J. (Eds). *Ecological evaluation of the Mirador-Río Azul National Park, Petén, Guatemala*. Wildlife Conservation Society. https://www.researchgate.net/publication/284730660_Ecological_Evaluation_of_Mirador-Rio_Azul_National_Park_Peten_Guatemala.

Belize Press Office. (2019, 25 July). Press release: *Ministry of national Security Comments on Sarstoon Incident*. https://www.facebook.com/GOBPressOffice/posts/press-releaseministry-of-national-security-comments-on-sarstoon-incidentbelmopan/2306836019351888/.

Bezaury, J., Cerezo, M., Gutierrez, D., Jacobs, N. & Mittermeier, C. (2005). The Mesoamerican Reef: A Shared Commitment for the Caribbean Jewel. In Mittermeier, R., Robles, G., Kormos, C., Mittermeier, C., & Sandwith, T. (Eds). *Transboundary Conservation: A new vision for protected areas* (pp. 313–319). CEMEX.

Boeglin-Naumovic, N. (2013). La denominada trocha fronteriza en la frontera entre Costa Rica y Nicaragua: breve análisis desde la perspectiva internacional. *LiminaR*, 11(2), 35–53.

Breaking Belize News. (2019, April 15). *Guatemalan navy gunboats block Belize Coast Guard from accessing the Sarstoon river*. https://www.breakingbelizenews.com/2019/04/15/guatemalan-navy-gunboats-block-belize-coast-guard-from-accessing-the-sarstoon-river/.

Brenner, J.C. & Davis, J.G. (2012). Transboundary conservation across scales: A world-regional inventory and a local case study from the United States–Mexico border. *Journal of the Southwest*, 54:3, 499–519.

Camargo, A. (2021). Stagnation. Waterflows and the politics of stranded matter in La Mojana, Colombia. In *Delta life: exploring dynamic environments where rivers meet the sea* (pp. 83–101). Berghahn Books.

Canty, S., Preziosi, R. & Rowntree, J. (2018). Dichotomy of mangrove management: A review of research and policy in the Mesoamerican reef region. *Ocean and Coastal Management* 157, 40–49.

Caribbean News (2023, October 8). *Caribbean Prime Minister cautions Belizeans against frequenting the Sarstoon*. https://caribbean.loopnews.com/content/prime-minister-cautions-belizeans-against-frequenting-sarstoon.

Carrillo, A., Ortega, A., Gama, L., Bello, E., Rioja, R. (2022). Effectiveness of management of the Mesoamerican Biological Corridor in Mexico. *Landscape and Urban Planning* 226, 104504.

Castro, E. (2022). Safeguarding ecosystem connectivity along and across borders. In Laako, H., Pliego, E., Ramos, D. & Marquez, B. (Eds). *The Guardians of the Maya Forest*. Discussion report published by the Journal *Versus*. https://www.versuslehti.fi/tiededebatti/the-guardians-of-the-maya-forest/ (Accessed in May 2024).

Cortesi, L. (2016). An ontology of water and land in North Bihar, India. *Journal of the Royal Anthropological Institute*, 27(4), 870–889.

Crooks, K. & Sanjayan, M. (2006). Connectivity Conservation: Maintaining connections for nature. In Crooks, K. & Sanjayan, M. (Eds). *Connectivity Conservation* (pp. 1–28). Cambridge University Press.

De la Torre, A., Camacho, G., Arroyo-Gerala, P., Cassaigne, I., Rivero, M., & Campos-Arceiz, A. (2021). A costeffective approach to mitigate conflict between ranchers and large predators: A case study with jaguars in the Mayan Forest. *Biological Conservation*, 256, 109066.

ECOSUR. (1995). *Protected Areas and Archaeological Sites—Mapping and database development to support conservation planning in the Selva Maya tri-national region.* Library, San Cristóbal de las Casas, El Colegio de la Frontera Sur, México.

Erg, B., Groves, C., McKinney, M., Michel, T.R., Phillips, A., Schoon, M.L., Vasilijevic, M., & Zunckel, K. (2015). *Transboundary Conservation: A systematic and integrated approach*. Best Practice Protected Area Guidelines Series, 23. International Union for Conservation of Nature and Natural Resources, Switzerland.

Finley-Brook, M. (2007). Green neoliberal space: The Mesoamerican biological corridor. *Journal of Latin American Geography*, 6:1, 101–124.

FAO, IUFRO & USDA (2021). A Guide to Forest-Water Management. *FAO Forestry Paper* 185. Rome. https://doi.org/10.4060/cb6473en.

FUNDAECO (2024). Río Sarstún Multiple Use Area, Guatemala. Fundación para el Ecodesarrollo y la Conservación. https://fundaeco.org.gt/fundaeco.org.gt/areas-trabajo/fichas-tecnicas/rio-sarstun.html

García, G. & Secaira, F. (Eds, 2006). *Una visión para el futuro: cartografía de las Selvas Mayas, Zoque y Olmeca.* Infoterra Editores, San José, Costa Rica, TNC.

García, R. & Radachowsky, J. (2004). Ecological evaluation of the Mirador-Río Azul National Park, Petén, Guatemala. Wildlife Conservation Society. https://www.researchgate.net/publication/284730660_Ecological_Evaluation_of_Mirador-Rio_Azul_National_Park_Peten_Guatemala.

Gómez-Pompa, A., Allen, M., Fedick, S., & Jímenez, J. (Eds, 2003). *The Lowland Maya Area: Three Millennia at the Human-Wildlife Interface*. CRC Press.

Herrera, A. (2004). *Sartoon-Temash National Park Management Plan.* Sarstoon Temash Institute for Indigenous Management, Punta Gorda, Belize. https://rris.biopama.org/sites/default/files/2021-02/Sarstoon%20Temash%20Management%20Plan%202004.pdf.

Iglesias-Prieto, N. (2021). Tijuana provocadora. Transfronteridad y procesos creativos. In Valenzuela Arce, J.M. (Ed) *Transfronteras: fronteras del mundo y procesos culturales* (pp. 97–127). El Colegio de la Frontera Norte.

Kauffer, E. (2019). Contrasting water securities: The Mexican state facing downstream stakeholders in the Suchiate transboundary river basin. *International journal of water resources development*, 35(1), 30–48.

Kauffer, E. (2020). Aguas y fronteras: encrucijadas de los territorios. *A&P Continuidad*, 7(12), 98–105.

Kauffer, E. (2024). Beyond conflict and cooperation in Central American transboundary river basins: Crossroads of actors and borders. In Batista da Silva, L.P., Costa Ribeiro, W., & Battistello Espíndola, I. (Eds). *New Perspectives on Transboundary Water Governance. Interdisciplinary Approaches and Global Case Studies* (pp. 61–80). Routledge. doi:10.4324/9781003333678-7.

Kauffer Michel, E.F. (2013). De las indefiniciones a las demarcaciones inacabadas: repensar las fronteras fluviales y terrestres entre México, Guatemala y Belice. *LiminaR*, 11(2), 70–81.

Kauffer Michel, E.F. (2018). Entre configuraciones geográficas y delimitación de fronteras: tres conflictos en cuencas compartidas de Centroamérica, África occidental y el sur de México. In Kauffer Michel E.F. (Ed). *Cuencas transfronterizas: la apertura de la presa del nacionalismo metodológico* (pp. 203–268). CIESAS.

Kauffer Michel, E.F. (2017). Entre rigidez política (hacia el conflicto) y fluidez hídrica (hacia la paz): Las fronteras de agua de México con Guatemala y Belice. *Revista de Paz y Conflictos*, 10(1), 61–86.

Keskinen, M., Guillaume, J.H., Kattelus, M., Porkka, M., Räsänen, T.A., & Varis, O. (2016). The water-energy-food nexus and the transboundary context: insights from large Asian rivers. *Water*, 8(5), 193.

Laako, H., & Kauffer, E. (2021). Conservation in the frontier: Negotiating ownerships of nature at the southern Mexican border. *Journal of Latin American Geography*, 20:3, 40–69.

Laako, H., Pliego Alvarado, E., Ramos Muñoz, D. & Marquez, B. (2022). Transboundary conservation and nature states in the Maya Forest: International relations, challenged . *Globalizations* 19:8, 1288–1310.

Latrubesse, E.M., Stevaux, J.C., & Sinha, R. (2005). Tropical rivers. *Geomorphology*, 70 (3–4),187–206.

Mittermeier, R.A., Robles Gil, P., Kormos, C.F., Mittermeier, C.G., & Sandwith, T. (Eds, 2005). *Transboundary conservation: A new vision for protected areas*. CEMEX.

Moreira-Ramírez, J.F., López, J.E., García-Anleu, R., Córdova, F., & Dubón, T. (2015). Tamaño, composición y patrones diarios de actividad de grupos de pecarí de labios blancos (Tayassu pecari) en el Parque Nacional Mirador-Río Azul, Guatemala. *Therya*, 6(2), 469–481.

Nations, J., Manterola, C., Rodríguez, C., & Hernández, R. (2005). The Maya Tropical Forest: Saving a Threatened Tri-National Biological and Cultural Heritage. In Mittermeier, R., Robles, G., Kormos, C., Mittermeier, C., & Sandwith, T. (Eds). *Transboundary Conservation: A new vision for protected areas* (pp. 147–157). CEMEX.

Niang-Diop, I., Dansokho, M., Ibrahima L.Y & Niang, S. (2002). Senegal National report. Phase 1: integrated problem analysis. Global Environmental Facility. https://www.scribd.com/document/152945754/1-Senegal-Report-country-wise.

Primack, R., Bray, D., Galletti, H., & Ponciano, I. (Eds, 1998). *Timber, tourists and temples: Conservation and development in the Maya forest of Belize, Guatemala and Mexico*. Island Press.

Ramsar. (2024a). Reserva de usos múltiples Sarstún. https://rsis.ramsar.org/ris/1667?language=es.

Ramsar. (2024b). Sarstoon Temash National Park. https://rsis.ramsar.org/ris/1562?language=es.

Romero & Cegarra (2024, February 15). This Century-Old Border Dispute Pits an Army Against Unarmed Volunteers. *The New York Times*. https://www.nytimes.com/2024/02/15/world/americas/belize-guatemala-border-dispute.html.

Sarstoon-Temash National Park Transcript of Stakeholders' Workshop (1998). Held on February 22, 1997 at the Community Center, Barranco Village, Toledo District, Belize. Belize Archives and Records Service.

Schroder, W., Murtha, T., Broadbent, E.N., & Almeyda Zambrano, A.M. (2021). A confluence of communities: Households and land use at the junction of the Upper Usumacinta and Lacantún Rivers, Chiapas, Mexico. *World Archaeology*, 53(4), 688–715.

Sehring, J., & Wolf, A.T. (2023). Affective Hydropolitics: Introduction to the Themed Section. *Water Alternatives*, 16(3), 900–911.

Seide, W.M., & Fantini, E. (2023). Emotions in Water Diplomacy: Negotiations on the Grand Ethiopian Renaissance Dam. *Water Alternatives*, 16(3).

Selva Maya (2019). *Aguadas y fauna asociada: Monitoreo de cuerpos de agua y fauna asociada en la Selva Maya.* Report by the Selva Maya Programme.

Selva Maya (2021). *Sistematización de la implementación y de los impactos de las líneas de monitoreo del proyecto Fomento del monitoreo de la biodiversidad y cambio climático en la región de la Selva Maya: Aguadas y fauna asociada.* Report by the Selva Maya Programme.

Syvitski, J.P., Cohen, S., Kettner, A.J., & Brakenridge, G.R. (2014). How important and different are tropical rivers? An overview. *Geomorphology*, 227, 5–17.

SRE (1893). Tratado sobre Límites con Honduras Británicas celebrado entre los Estados Unidos Mexicanos y el Reino Unido de la Gran Bretaña e Irlanda (Belice), SRE. http://www.ordenjuridico.gob.mx/TratInt/2022bl/BELICE-LIMITES.pdf.

Talley, D., Huxel, G., & Holyoak, M. (2006). Connectivity at the Land-Water Interface. In Crooks, K. & Sanjayan, M. (Eds). *Connectivity Conservation* (pp. 97–129). Cambridge University Press.

The San Pedro Sun (2022, September 19). *Guatemalan incursion on the Belize side of the Sarstoon River.* https://www.sanpedrosun.com/community-and-society/2022/09/19/guatemalan-incursion-on-the-belize-side-of-the-sarstoon-river/.

TIES (2015). Definition of Eco-Tourism. The International EcoTourism Society. https://ecotourism.org/what-is-ecotourism/.

Valenzuela Arce, J.M. (2021). Transfronteras y límites liminales. In Valenzuela Arce, J. M. (Ed) *Transfronteras: fronteras del mundo y procesos culturales* (pp. 17–42). El Colegio de la Frontera Norte.

Waterbury, J. (1979). *Hydropolitics of the Nile valley.* Syracuse University Press.

WCS (2022). *News From COP27 Climate Conference: European Union announces EUR 25.5 M investment in Mesoamerica's Five Great Forests.* Press release. https://newsroom.wcs.org/News-Releases/articleType/ArticleView/articleId/18267/News-From-COP27-Climate-Conference-European-Union-announces-EUR-255-M-investment-in-Mesoamericas-Five-Great-Forests-English-and-Spanish.aspx (Accessed in May 2024).

Zbicz, D. (2003). Imposing Transboundary Conservation. *Journal of Sustainable Forestry*, 17:1–2, 21–37.

4 Maya Forest Waterlands as Waterless Transboundary River Basins for Their Inhabitants

Edith Kauffer

Introduction

As defined in Chapter 1, waterlands conceived as mobile confluences of waters and lands with open-ended boundaries suppose the abundance of waters that, in some cases (as in floodplains or lowlands or riversides presented in Chapter 3), convert into an excessive presence when overflows arise. The reality of the abundant water that characterizes the Maya Forest waterlands is related to climate variability along the year; it is typical of tropical regions, which are defined by the alternance of wet and dry seasons, and sometimes, it means extreme situations, such as flooding or drought. Consequently, when compared with other places of Mexico and Guatemala that are naturally water-scarce, this waterland evidences a natural abundance of waters and a high availability per capita, according to data (Kauffer, 2019). Nevertheless, even in normal cycles of seasons, water abundance does not necessarily reflect in the local population's daily life in terms of access to public water services and sanitation. The situation of remote borderlands and peripheral regions constraints access to water for human consumption as well as adequate conditions of sanitation, even in areas with plenty of water. This chapter focuses on this deep contradiction in the context of a natural, geographic, but also socially and politically produced reality of these "human-water-land systems" (Camargo, 2022), which are also known as transboundary river basins (TRBs).

The notion of river basins has been historically based on a natural, geographic, and hydrographic delineation born in the eighteenth century in France and proposed by Buache (Ghiotti, 2006) around the confluence of surface waters that flow to a common exit, which is delimited by the watershed, or line, joining the points of maximum altitude. The utopic control of large rivers through hydraulics during the eighteenth, nineteenth, and part of the twentieth centuries directly promotes the management of the river basin. One of the most recognized examples of this narrative has been the Tennessee Valley Authority at the beginning of the 1930s in the United States (Kauffer, 2014). Furthermore, the river basin became associated with the Integrated Water Resources Management (IWRM) approved by the Dublin Conference in 1994 and is still considered a pillar of water policies at the international scale (Kauffer & Maganda,

DOI: 10.4324/9781003429050-5

2022). Although the dominant conceptualization of river basins by researchers has tended to focus on the river basin as a natural delineation, from their appearance in the eighteenth century to the present times, river basins have always been political spaces used by governments and other stakeholders to pursue diverse goals. Deciding upon the limits of a river basin and its name are, above all, matters of political choices and decisions.

What Ghiotti (2014, p. 11) designates the "internationalization of water management" is marked by the emergence of the concept of transboundary river basin (TRB) during the 1990s in the fields of both research and policy. Nevertheless, despite its apparent focus on hydraulics and hydrology and its separation from administrative boundaries, the river basin, when adopted across borders, signifies a politicization process (Ghiotti, 2014). The water management framework, once centered on state sovereignty (through international watercourses), changes and includes water sharing, and hence threats to the state's sovereignty when focusing on TRB.

The transboundary river basins (TRBs) propose a distinct form of interaction between political borders and waters (Kauffer, 2024), in addition to the other ones presented in Chapter 3. They profoundly connect both elements, and also land, natural resources (in our case, forests), and the complete lives of local people in the territory, as the "forest people" discussed in chapters 2 and 5. TRBs are hence a form of entanglements between water, land, and forest, as well as a type of delineation of space that is different from those mentioned in Chapter 1 – namely, ecoregions and hotspots.

This chapter contributes to analyzing the Maya Forest waterlands by focusing on the contrast between these waterlands of plentiful water and the realities of domestic uses of water and sanitation by the local population, as well as their relations with political borders. Does the location on one side of a border or another matter, or does the transboundary situation prevail? Does the TRB configuration regarding the border and surface waters create differences? Apart from sharing waters and forests, do the local populations in the studied borderlands have common conditions of access to water and sanitation?

The chapter outlines three main points: (1) a general conceptualization of TRBs, including the context of the study region; (2) a characterization of the two TRBs as sub-basins, their relations with the borders, focusing on the abundance and diversity of water bodies, forests and protected areas (PAs); and (3) the shared and contrasting reality of drinking water and sanitation services for local inhabitants in both TRBs. The analyzed data demonstrate a lack of water as a result of shared social and political conditions along the Maya Forest waterlands, but they also evidence differences according to the country. This chapter suggests that the abundance of water in the Maya Forest waterlands reveals the existence of plenty of water TRBs, although this does not result in water availability for the local population; presents huge, shared water challenges; and contrasts with the reality of waterless TRBs in the context of waterful borderlands.

Transboundary River Basins (TRBs): A Recent Concept and Its Realities in the Maya Forest Waterlands

Transboundary river basins are sometimes designed as "international river basins" or "shared river basins". Although those terms might appear to be synonyms in the common language, they denote contrasting meanings according to political contexts. Fieldwork experiences in Mexico and Central America evidence that the vocabulary regarding transboundary water issues is not neutral because the recognition of their existence is generally not welcome by national government officials, which will be explained below in more detail. Consequently, the use of this terminology has potential political implications regarding each of the three notions of "transboundary river basin", "international river basin", and "shared river basin" (Kauffer Michel, 2011).

"Transboundary river basin" expresses the politically most neutral term, and therefore, it can be employed more comfortably by states because it only alludes to the geographical reality that the river basin extends beyond one or more political borders. It merely indicates a technical fact that has no consequences for states' activities and engagements. In contrast, the notion "international river basin" implies that, by extending beyond the limits of any single nation-state, the basin may be the target of international cooperation that supersedes and threatens national sovereignty (Allouche, 2010, p. 57; Lasserre & Boutet, 2002). The third notion, "shared river basins", suggests and implies cooperation on "common" resources. It is mostly used by nongovernmental organizations (NGOs), such as the International Union for Nature Conservation (IUCN) (Aguilar and Iza, 2009). It is almost never used by states and is only occasionally used by academics (e.g., Mirumachi & Chan, 2014; Olvera, Kauffer & Schmook, 2011; Zeitoun, Goulden & Tickner, 2013). Commenting upon the term "shared river basins", Cascão & Zeitoun (2010, pp. 34–35) assert that "not all transboundary waters are shared" as the notion of sharing necessarily implies equity that is not always present in transboundary water issues.

Additional cautionary notes about terminology are found in the work of Sneddon and Fox (2006), who highlight that the political notion of *basin* is controversial for the Mekong TRB, which concerns six Asian countries. Its mere use seems to imply a political stance in favor of international cooperation, which automatically means positioning a threat to state sovereignty. The same situation occurs with the absence of explicit reference to TRBs in the texts of the 1997 United Nations Convention on the Law of the Non-Navigational Uses of International Watercourses and of the 1992 Convention on the Protection and Use of Transboundary Watercourses and International Lakes of United Nations Economic Commission for Europe (UNECE), although the latter has modified the terminology. The simple notion of TRB is not consensual; thus, as explained by Ghiotti (2014) about the international literature, between 1991 and 2014, natural and environmental sciences mostly used other notions, such as *international waters, transboundary waters*, and *transboundary rivers* to avoid TRB.

Hence, when a river basin extends beyond national boundaries, the use of terminology is crucial and highly controversial (Allouche, 2010). Consequently, the concept of TRB is troublesome for international relations and the academic literature owing to its political ontology and threats to the state's sovereignty. In the former venue, the notion has been bypassed through strenuous states negotiations, while in the latter, it is either not defined or is referred to according to biophysical traits.

In the first years of its appearance, the notion of TRB was used in an undifferentiated way by various disciplines in international literature (Ghiotti, 2014). The dominant perspective of the definition of *river basin* centered on the physiographic dimension, which was also used in defining the notion of TRB – and all the more so because of its eminently political character, which implied a threat to states' sovereignty and a requirement for depoliticization.

Consequently, defining a TRB based on its biophysical components, or border-crossing general features, is the most common position (Teasley & McKinney, 2011; Mirumachi, 2015). The International Network of River Basins Organisations (INBO) and Global Water Partnership (GWP) (INBO-GWP, 2009, p. 12) define TRBs as "basins that cover more than one country". The TWAP specifies that TRBs are "rivers for which the hydrological boundaries cross an international border, even by a relatively small amount" (UNEP-DHI & UNPE, 2016, p. xii). The academic literature – even those articles whose core issue is TRBs – also abounds in works that either give only general definitions of TRBs or leave them undefined.

An additional practice that impedes debate on the notion of TRBs is the frequent use of different notions as synonyms. For example, Milich and Varady (1999) employ "international river basins" as synonymous with TRBs. Caponera (1993) similarly uses the phrase "transboundary rivers", just as other authors refers to "international rivers" (Sadoff & Grey, 2005) or "international watercourses" (Waterbury, 1997). Despite such impediments, and although it is the most neutral reference, the deep political component of TRBs is increasingly mentioned through reference to upstream-downstream positions (Milich & Varady, 1999) and other political issues.

From a purely biophysical perspective, TRBs are fixed and naturally defined by the watershed line that joins the highest points in altitude around a merging of rivers and surface waters. However, when their political and social components are considered, TRBs are more than drainage basins organized around the confluence of rivers and other connected water bodies, which flow to an exit through a delineated watershed. They are also territories produced by power relations in their definition, name, limits, and transboundary relations. Moreover, the main element of differentiation between populations that live in a shared TRB is the existence of one or more international border. Living on one side of the border or on the other may be highly relevant in terms of access to infrastructure and water services.

TRBs are defined according to water confluences that entangle with political borders, whereas waterlands refer to the water–land nexus. In both cases, the notions of TRBs and waterlands refer to cross-boundaries realities, the former inviting to go beyond political borders and the latter transcending the Western ontological separation between water and land. Both TRBs and waterlands are thus connected with boundaries, since they propose to outstrip these more or less porous lines of division. Owing to the type of physical limits that define them from a traditional perspective, TRBs are more fixed than waterlands, the boundaries of which are porous and fluctuating. However, as political territories, TRBs are also defined according to human relations and, among them, power relations and stakeholders' interests. In this sense, TRBs may also be shifting, and their delineation, name, and spaces of management depend on national institutions, interests, sovereignty, and, above all, the capacities to establish transboundary cooperation initiatives.

TRBs and waterlands constitute two interconnected perspectives of analysis that entangle water and land, although from diverse standpoints. TRBs follow waters that transcend the borders to include and understand land, forest, and environment issues along a water-defined space or territory of connections. However, waterlands may focus on the same components, but try to account for their fluidity, hybridity, and constantly transforming shapes. Owing to its fixed characteristics, once its boundaries have been delineated, only the TRB enables a precise quantification of population and access to water services and sanitation (as discussed in this chapter) to understand the importance of the political border for the daily life of the riparian population in relation to waters and forest coverage.

In the case of the three borders considered throughout the Maya Forest waterlands – the Mexico-Guatemala border, defined by the 1882 Limits Treaty; the Belize-Mexico border, delineated by the 1893 Treaty between the latter and British Honduras; and the disputed "adjacency line" between Belize and Guatemala – the TRBs have not been jointly delimited by states, nor mapped. This situation explains the lack of an official and interstate map of TRBs in the region (Kauffer, 2019). River basins maps shaped from a strict national perspective do not coincide when each state delimits them. This is the case between Mexico and Guatemala owing to hydrography, geography, topography, and the size of national territories and TRBs. When the Mexican and Guatemalan national river basins maps are "pasted", the river basin limits do not coincide; it is thus necessary to delineate them by considering the whole TRB.

Moreover, the realization of a map of TRBs depends on the source and availability of information and access to it in each country, which is also a key international principle defined by the 1997 United Nations Convention on the Law of the Non-Navigational Uses of International Watercourses. However, sharing information on waters and rivers in Central America is deeply complex when states consider them national security issues and do not wish to share them with foreigners. Handing over information to states or individuals that may have researchers of bad territorial intentions is not a common practice. As a researcher, I never reached official access to the geographical data of the governments of Guatemala and Belize. Guatemalan authorities clearly denied access, and authorities in Belize gave me a CD with the information available on the Internet but not the precise geographical information formally requested and authorized for its sharing, which the CD was supposed to contain. The number and delimitation of TRBs – especially in downstream Mexico and Belize in lowlands and floodplains – is complex and has given diverse results, according to different authors (Kauffer Michel, 2018). For that reason, it requires precise information in GIS shapes.

Following the delimitation of García García & Kauffer Michel (2011) and Kauffer (2024), from a Mexican perspective, Mexico, Guatemala, and Belize share six TRBs: four of these between Guatemala and Mexico (Suchiate, Coatán, Grijalva, and Candelaria) and two that are trinational (Usumacinta and Hondo) (Map 4.1). According to the map presented in the introduction that delineates the Maya Forest, three of the six TRBs extend in the Maya Forest waterlands. From the Caribbean Sea to the Gulf of Mexico, these TRBs are the following: the Hondo River basin, between Guatemala, Belize and Mexico; the Candelaria River basin, between the Petén Department in Guatemala and the state of Campeche, in Mexico; and the Usumacinta River basin, which is the focus of this chapter.

The Belizean Mexican border, as described in Chapter 3, is mostly defined by the Hondo River that also shapes the Hondo TRB. As in other cases, its delimitation may be different according to the sources and interpretation of confluences (Kauffer, 2021). Map 4.1 indicates that the Hondo TRB is located upstream in Guatemala (30.3%) and downstream between Mexico (51.7%) and Belize (18%). The complex delimitation in floodplains, owing to the difficult access to precise, small-scale geographical information for Belize, has been a point of controversy among researchers, and the lack of an interstate-agreed map explains the difference of limits in the maps produced by researchers.

Along the border or adjacency line between Belize and Guatemala, the TRBs are three: Belize, Moho, and Temash rivers. Although a joint delineation map does not exist owing to the political complexity of the upstream (Guatemala)-downstream (Belize) dynamics in a context of border dispute, the delineation of coastal river basins is simplest to establish when the downstream section is located in the same state. Eventually, the Sarstoon TRB that includes the fluvial line between both countries, although highly complex from a political perspective, does not register problems of limits owing to its location on a corridor of coastal river basins.

As an area defined by this book, the Maya Forest waterlands contain seven TRBs, the extension of which are distinct and present diverse water issues. They cover the three defined waterlands corridors mentioned in the introduction, parts of the three states (Belize, Guatemala, and Mexico), and the three studied borderlands coined in Chapter 3. Each of these TRBs also shapes one or more forest waterlands where waters entangle in diverse manners with land and forest, but also where they cross international borders (and sometimes delineate them), including protected areas, confluences of rivers, and areas of wetlands. These dynamics confer, to the TRBs, a mix of natural and sociopolitical entanglements that are necessary to understand in order to analyze them as waterlands with plenty of water.

The chapter presents the results of twenty years of qualitative research in the region and proposes two specific focuses of analysis: it deals with the major TRB in extension and water resources – namely, the Usumacinta River basin and two Usumacinta's sub-basins located in two distinct points of the Maya Forest waterlands (the Gran Usumacinta River sub-basin and the Santo Domingo River sub-basin[1]) – which means diverse conditions regarding waters and borders. These two case studies are associated with official data from Mexico and Guatemala's census of 2000, 2002, 2010, 2018, and 2020 and completed with quantitative data obtained during fieldwork

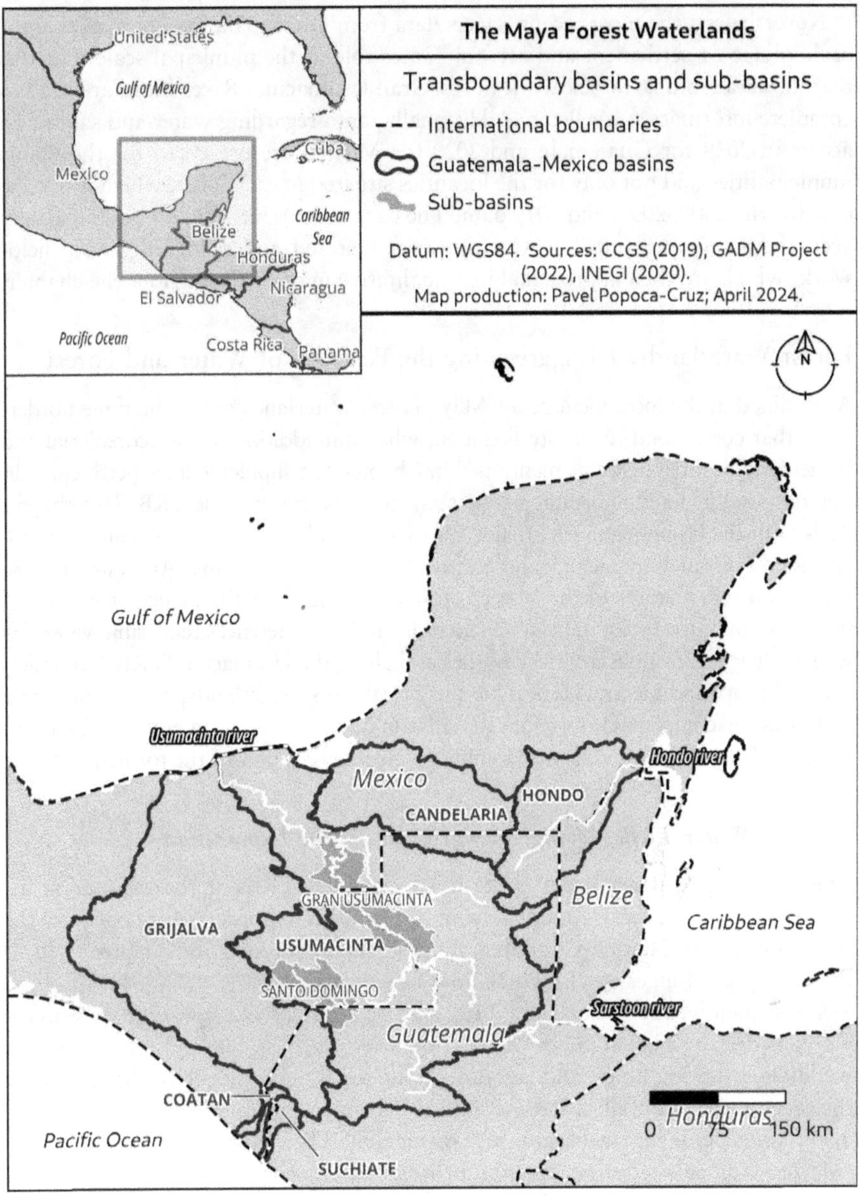

Map 4.1 Transboundary Basins and Sub-basins in the Maya Forest Waterlands
Source: Pavel Popoca Cruz

in 2018 in Guatemala. In Guatemala, the census was conducted in 2002 and 2018, but information about the last census was published at the end of 2019. The former census does not present information availability for the Petén department. The Mexican census is conducted every 10 years: 2000, 2010, and 2020.

Nevertheless, the most recent census data from Guatemala are not available for each village or settlement and are only accessible at the municipal scale. For this reason, data from 2018 fieldwork in the Gran Usumacinta River basin are used to complete information for Petén. Additionally, data regarding water and sanitation access in 2018 for Guatemala and 2020 for Mexico are presented for the whole municipalities and not only for the localities situated in each TRB, which is not the case for the 2000, 2002, and 2010 data. The data are not intended to be statistically accurate but to illustrate the situations encountered and observed during fieldwork, which are then approached in a qualitative manner throughout the chapter.

Forest Waterlands? Disaggregating the Features of Water and Forest

As outlined in the introduction, the Maya Forest waterlands are in the three borderlands that correspond to the studied area, where, in addition to the ecoregional and biogeographical delineations mentioned in Chapter 1, a supplementary pertinent delimitation that includes and entangles waters, land, and forest is the TRB. This chapter deals with the Usumacinta (the major river basin) and two of its transboundary subbasins, the Gran Usumacinta and Santo Domingo River basins. Analysis of these focuses on two features of the Maya Forest waterlands: the abundance of water and the situation of the forest milieu. The focus is on their differences regarding waters, as well as their location vis-à-vis the border and along the Usumacinta TRB. To understand the apparent contradiction of these waterless waterlands, the chapter first addresses the complete Usumacinta River basin before focusing on both sub-basins of the Gran Usumacinta and Santo Domingo watersheds as part of the former.

Plenty of Water TRBs: Living among Waters in the Usumacinta

The Usumacinta River basin is the most extended TRBs at the regional scale, considering all Central America from the southern border of Mexico to the south until the Panama-Colombia border. As mentioned in Chapter 3, it is crossed by the longest and highest flow rate.

Consequently, the Usumacinta TRB may be considered a space full of water if we consider the high number of its tributaries, such as big and small streams, wetlands, watering holes, and lagoons. Along the Usumacinta River basin, during the wet season, rainfall is intense, and the average precipitation can reach over 4,500 mm per year upstream in Guatemala. The river basin is trinational (Mexico-Guatemala-Belize), but the Belizean part is tiny – less than 0.1% of its extension – and this section has neither population nor surface water resources. Consequently, the main dynamics of the river basin are binational.

Despite numerous historical dam projects, the Usumacinta River basin in Mexico is free of dams. In upstream Guatemala, the Chixoy river is dammed by the biggest Guatemalan infrastructure built at the beginning of the 1980s and imposed by the government to Indigenous local population by force (Kauffer Michel, 2013). Another upstream affluent, the Xalbal River, also has two hydropower infrastructures that began operating in 2010 (Hidro Xacbal) and 2017 (Hidro Xacbal Delta).

The Usumacinta TRB extends over 77,225 km^2, the most extended portion of which is in Guatemala (55.8%), the catchment headwaters (Map 4.1). In Guatemala, the basin extends over part of the departments of Huehuetenango, Quiché, Alta Verapaz, and Petén. In Mexico, the river basin covers portions of the states of Chiapas (upstream and midstream), and Tabasco and Campeche (downstream). The whole Usumacinta TRB is divided into seventeen sub-basins; among them, one is in Guatemala; eight are in Mexico, and eight are also transboundary sub-basins. Among these transboundary watersheds, we find the following: Chixoy, Chajul, Ixcán, Santo Domingo, Comitán, or Pojom (depending on each national perspective), Upstream Lacantún, San Pedro, and Gran Usumacinta. The largest transboundary sub-basins correspond to the main tributaries, the Chixoy and San Pedro rivers. I have selected two transboundary sub-basins for this study to analyze the situation of both countries regarding waters and forest and to understand the role of the borders in the context of TRBs. This section deals with the description of each selected transboundary sub-basin, their characteristics regarding watercourses, and the borders; then, it is crucial to eventually analyze how these waterful river sub-basins are also waterless realities for local populations and how the political border involves differences in access to water-related services.

Image 4.1 The Usumacinta River Flowing Downstream with Mexico on the Left Riverbank and Guatemala on the Right One
Source: Edith Kauffer, August 2022

The Gran Usumacinta Sub-Basin[2]: A River and Three Borders

As a portion of the Usumacinta River basin, the Gran Usumacinta transboundary river sub-basin is located both in the north of the Lacandon Forest waterlands and in the south of the Laguna del Tigre-Gulf Riverlands corridors. As indicated by its name, which calls to mind its centrality for the whole TRB, the Usumacinta River flows through this sub-basin, upstream as part of the international border and then downstream as the delineation of the subnational border between Chiapas and Tabasco until the head of the municipality of Balancán, in Mexico. The Gran Usumacinta sub-basin initiates upstream at the confluence of the Lacantún and Usumacinta rivers, which is a lowlands area located on the border (see Chapter 3) and following the river to the north. It extends to the inaccessible part of the watercourse and its rapids, from the Piedras Negras archaeological site to the state of Tabasco. Farther and downstream, it also includes the Boca del Cerro Canyon and the floodplains of the municipalities of Tenosique and Balancán, in Tabasco. Thus, the Gran Usumacinta is characterized by a double transboundary situation around the international watercourse: one at the international scale and the other at the subnational scale, between Chiapas and Tabasco.

Regarding water and climate, the Gran Usumacinta sub-basin has a warm, humid climate that fluctuates between 22° and 28° C annually, with a warmer part in the Guatemalan inland (25°–28°) and downstream in Tabasco, Mexico (26°–28°); and a slightly colder one (22°–24°) following the watershed line in Mexico, which is situated in the west of the TRB. The rainfall reaches 1,500–1,599 mm a year in the main part of Guatemala and in Mexico between 2,000–2,500 mm upstream and downstream and the major quantity of rain between 2,500–3,000 mm that corresponds to the upstream Mexican portion. Consequently, the Mexican part of the Gran Usumacinta receives the most rainfall and all the flows from the whole basin – that is, the northern Mexican floodplain.

The Gran Usumacinta sub-basin extends over two municipalities of Guatemala, La Libertad, and Las Cruces, both located in the Petén department. Las Cruces is a new municipality created in 2011 and separated from La Libertad at the request of the local population. The name Las Cruces (The Crosses) was proposed by the inhabitants and may have a double meaning. The referred crosses could indicate a territorial mark left by historical actors from the Maya Forest waterlands, such as looters of archaeological sites, according to the official history; or, left by *chicleros,* the gum collectors, analyzed by Laako in Chapter 5. The creation of new municipalities is a typical phenomenon of frontier areas and illustrative of a type of borderlands. It has also been the result of national strategies for marking the political border of both states by colonizing remote areas, in the nineteenth and twentieth centuries before and after the Limits Treaty of 1882 (Torras, 2019; Rodas, 2014), with new population settlements and municipalities. Today, this process is more complex owing to the diverging interests of different stakeholders.

The municipality of Las Cruces extends along the Usumacinta River, and the municipal capital does not figure as part of the Gran Usumacinta sub-basin as it is located inland from the riverside. A total of 11,618 people living in 20 villages and small settlements such as hamlets, farmhouses, or agricultural endowments (with only five localities with more than 1,000 inhabitants among them) are part of the river basin, which is a rural and remote area according to fieldwork in 2018. The main village is Bethel, with 1,687 inhabitants in 2018, which is also the only official border-crossing point on the Guatemalan riverside of the Usumacinta for national people, international tourists visiting the Maya Forest route, and conservationists and archaeologists.

Inland from the riverside is La Libertad municipality, with 41 villages and hamlets in the sub-basin that do not include a municipal capital. They are more numerous but smaller and more remote than in Las Cruces as this corresponds with the Sierra Lacandón National Park, which borders Mexico on the west, in the Usumacinta River, and on the north with the terrestrial line that separates Petén from Tabasco. La Libertad counts 49 villages, with five settlements with more than 1,000 inhabitants, for a total of 9,936 inhabitants in 2018 in the Gran Usumacinta River basin, according to fieldwork.

Both Las Cruces and La Libertad represented about 18% of the Gran Usumacinta total population in 2018 and only 11% in 2002; thus, the Guatemalan part of the sub-basin was and still is the most rural of the whole area. It is also notable that except for the Usumacinta River, no other major river is located on the Guatemalan section of the sub-basin, despite the numerous waterholes in the forest. An interesting fact is that some of the villages located in the sub-basin are cooperatives created during the 1960s as part of a state policy to colonize the Petén department through the Petén Promotion and Development Company (Empresa de Fomento y Desarrollo de Petén, FYDEP) and as a result of the return of Guatemalan refugees settled in Mexico from the mid-1990s. In both cases, those were political processes that included colonizing the area by creating new settlements in forested areas for peasants.

Although the Guatemalan side of the Gran Usumacinta River sub-basin corresponds to the Sierra Lacandón National Park, the Mexican part of the basin encompasses the opposite bank of the international river in the south, and includes the whole basin in the north, owing to the land border that divides Guatemala from Mexico horizontally (Map 4.1 and Image 4.2).

The Mexican riverbank belongs to the municipalities of Ocosingo and Palenque in the south of the Gran Usumacinta. In both cases, as well as on the Guatemalan riverside, the capital of the municipalities is located out of the basin. Nevertheless, the Mexican side is more accessible by road, especially since the 1994 Zapatista uprising that provoked the building of the bordering paved road, which was finished in 2000, following the gravel road previously established in the 1980s. Both the so-called Bordering Road (*Carretera Fronteriza*), which follows the river from north to south, and the southern land border have provoked colonization and deforestation processes in the whole region, although they have facilitated the local population's access to better

Image 4.2 The Gran Usumacinta River Sub-basin, Picture Taken from Mexico with the Lacandón National Park on the Opposite Riverbank
Source: Edith Kauffer, January 2018

living conditions. Consequently, the number and size of Mexican settlements are higher than those of settlements in the Guatemalan's riverbank, representing 19.3% of the Gran Usumacinta total population, with 21,554 inhabitants in 2018. Two villages are the most populated: one Indigenous village situated on the Mexican riverside, with more than 5,000 inhabitants, and the second one located on the road that reaches the capital municipality of Palenque riding to the north, with more than 1,000 people (Image 4.3).

In the north and downstream, the Gran Usumacinta sub-basin corresponds with the floodplain of Tabasco, which consists of flood zones for livestock farming with plenty of water, especially during the wet season. In some villages located on the riverside, inhabitants have two houses: one on the riverbank and another one located on natural hillsides when the river arises. Three municipalities of Tabasco are concerned: Emiliano Zapata, with a small portion in the Gran Usumacinta (1.8% of inhabitants); and Balancán and Tenosique, both with their municipal capitals in the sub-basin, two towns of more than 10,000 inhabitants. Tenosique represents 43.7% and Balancán 10% of the total population of Gran Usumacinta as they include urban areas.

Consequently, the total balance between rural and urban population in the Gran Usumacinta in 2018 was 64% versus 36%. It is worth noting that the

Image 4.3 The Gran Usumacinta River Sub-basin, Downstream in Tabasco, Mexico
Source: Edith Kauffer, December 2021

population of the Guatemalan part increased by 88% from 2002 to 2018 and between 21% (Ocosingo) and 27% (Palenque) in Chiapas, shaping a deep colonization process through rural and most remote areas. The other municipalities only registered a low population rise compared with the former: 9.8% in Balancán; 6.7% in Tenosique, and 5.4% in Emiliano Zapata. Moreover, Gran Usumacinta was, above all, characterized by 172 villages with less than 100 inhabitants in 2002 and 162 with less than 1,000 in 2018, that is, a situation of dispersed population.

The Gran Usumacinta River sub-basin shapes a world full of water – a region crossed and often flooded by a major river, which also forms part of the border between two states, Mexico and Guatemala, and a waterlands where waters and lands often merge and the tropical forest emerges, although grasslands have replaced tropical forest in some parts such as Tabasco. It is thus a colonization area and a borderland characterized by three border situations: the Usumacinta fluvial border, the land boundary between Guatemala and Mexico, and the subnational border between Chiapas and Tabasco.

The Santo Domingo Transboundary Sub-Basin: A Meandering River

The main watercourse of the Santo Domingo River basin has its source in the Chiapas hinterland and flows until it reaches the land border located in the south. Close to the border, the river merges into Guatemalan tributaries and then flows back to the interior of Chiapas before going back to the border, where it delineates a tiny fluvial portion of the line and enters Guatemala, quickly flowing back another time to the Mexican inland. Furthermore, it receives various Mexican tributaries until the confluence of the Jataté River that runs from the Cañadas area, well known as a Zapatista bastion from 1994. Downstream of this confluence, its name changes into Lacantún, a major affluent of the Usumacinta River mentioned in Chapter 3. The main part of the sub-basin is in Mexico, but the Guatemalan portion is important regarding the flow of upstream waters and abundance of rainfall. The river is not navigable except small portions for local transportation to the opposite riverbank as it runs through mountainous areas and a rugged topography with numerous waterfalls. The villages located along the river basin in both countries are small, numerous, and parsed, with 105 in Guatemala and 464 in Mexico, for a total of 569 in 2010.

The San Domingo TRB is more diverse in terms of climate and precipitation than the Gran Usumacinta, and it presents from warm humid to temperate subhumid climates, depending on the topography and elevation. For the same reasons, the annual average precipitation also varies from 800 mm to 4,000 mm in

Image 4.4 The Santo Domingo River Sub-basin, Picture from Mexico with Guatemala in the Background
Source: Edith Kauffer, March 2022

Mexico and from 1,400 to 4,899 mm in Guatemala. This is a rainier river basin than that in the Gran Usumacinta, but the main river is smaller, owing to the mountainous topography. In upstream and midstream Guatemala, the average temperature fluctuates between 8° and 25°, and in Mexico, located upstream at lower altitudes, midstream, and downstream, it oscillates between 14° and 26°.

In the Mexican portion, the major settlement is the capital of the municipality of Las Margaritas, with 20,786 inhabitants in 2010 (24,326 in 2020), and the village of Plan de Ayala with 3,164 in 2010 (3,804 in 2020). In Chiapas, the Santo Domingo TRB extends over five municipalities: Las Margaritas, with 70% of the population in 2010; La Trinitaria, with 5.3%; and Maravilla Tenajapa, with 4.2%, all bordering municipalities. Two are hinterland municipalities: Altamirano, which only concerns a tiny portion in the Santo Domingo TRB, with only 0.3% of inhabitants; and La Independencia, with 7.3%. Moreover, in 2010, 87.1% of the Santo Domingo population was living in Mexico, whereas 12.9% was settled in Guatemala.

Maravilla Tenejapa is a municipality created in 1999 in Chiapas as part of a process of remunicipalization to face the formation of autonomous rebellious municipalities by the EZLN in 1997–1998. In July 1999, seven new municipalities were created by the Chiapas governor and approved by the local congress as an imposed counterinsurgency strategy but also as a negotiation with local powerful groups from the new municipalities, despite the opposition of some other organizations (Leyva & Burguete, 2007). The creation of the new municipality tried to displace the Zapatista hegemony in the region (Leyva Solano & Rodríguez Castillo, 2007).

In Guatemala, the villages of the Santo Domingo TRB are smaller, less numerous, and distributed in three bordering municipalities located in the Chuchumatanes highlands. Their population presented the following data in 2010: Nentón, 1.8% of the Santo Domingo TRB's inhabitants; Barillas, 4.7%; and San Mateo Ixtatán, 6.4%. Some of them located close to the border were created during the last decades, owing to internal displacements and the repatriation of Guatemalan refugees settled in Mexico.

Bulej was and is still the main populated place in the Santo Domingo TRB in Guatemala with 2,530 inhabitants in 2002 and 9,865 in 2020, which is a steep increase. The municipalities of Guatemala are characterized by small and dispersed villages but increasing populations. No head of municipality is located in the Guatemalan Santo Domingo TRB, and the villages are remote, with difficulty of access and problems of public services, as the fieldwork evidenced.

In total, 198 villages had less than 1,000 inhabitants in 2000–2002 (218 in 2010) and 325 less than 100 people (332 in 2010). In both countries, population growth in the Santo Domingo River sub-basin is high, although it is more elevated in Guatemala than in Mexico.

The Santo Domingo sub-basin is a mountainous meandering river that receives flows from many tributaries on both sides. Its topography has induced governmental hydropower projects during the last decade that have provoked conflicts at the local scale, impeding their realization in Mexico but with major episodes of violence and unresolved situations on the Guatemalan side.

In both TRBs, we observe differences in climate and rainfall as well as the configuration regarding the main rivers and borders. The borderland situations also evidence deep differences due to remoteness and difficulties of access that are more marked in Guatemala, whereas the accessibility of both river basins is now easier, owing to infrastructures that sometimes have counterinsurgency purposes. Furthermore, both TRBs were and still are areas of colonization promoted by the states in the past and part of local and national population dynamics today. As borderlands, they share histories of political and social conflicts, as well as interests in controlling the borderland territories as peripheral areas. Hydropower plans were and still are major water issues in both sub-basins with plenty of water. However, their remoteness from political centers has allowed tropical forest vegetation to persist in some cases in small or large extensions.

Forest TRBs? Soil Uses and Protected Areas (PAs) in Borderlands

This section deals with the forest component of the Maya waterlands in both TRBs. The Gran Usumacinta and Santo Domingo TRBs are rather different in this aspect. To understand their distinctiveness, we must examine their contrasts regarding soil uses, as well as their relations with the presence of protected areas (PAs) and with the international borders. Sharing a TRB means accessing common water resources, but it is also rooted in a perspective of cooperative management of both waters and vegetation. In this sense, the joint management of TRBs should seek to equitably distribute waters and also manage the whole river basin, considering environment, biodiversity, and, in our area, the presence of forest. Each TRB presents a distinct situation according to the colonization history, extension of the agricultural frontier, and physical characteristics and topography. During the last three decades, the loss of forest coverage has been a deep phenomenon that has converted the whole Usumacinta TRB into a "deforestation hotspot" (Gallardo-Cruz et al., 2021, p. 97). Whereas the tree coverage represented 75% of the TRB in 2000, the forest decreased 27% in 18 years, and the differences between forest in PAs and outside PAs are evident (Gallardo-Cruz et al., 2021). In Mexico, forest loss in PAs represents 8% and 24% outside them. In Guatemala, forest loss has been higher and totally disparate: 37% for protected forest and 29% outside PAs (Gallardo-Cruz et al., 2021). Consequently, for presenting the characteristics of each transboundary sub-basins, it is interesting to consider PAs, because they do not imply the same consequences in each country. Apart from dividing states, the political border also defines differences in terms of the effectiveness of PAs.

The Gran Usumacinta TRB presents a huge contrast between the south, located along the river, and the northern floodplain of Tabasco. The land boundary that separates Guatemala from Mexico in the north also shapes a natural separation between tropical forest and grassland. Thus, the Guatemalan section of the Gran

Usumacinta has the most forest coverage, owing to the Sierra Lacandón National Park, which has contributed to its conservation, but also to social conflict, due to agrarian pressures and contemporary colonization processes in the Petén.

Six protected areas correspond to parts with the most forest coverage of the Gran Usumacinta River basin. Almost all La Libertad municipality and part of the Guatemalan Usumacinta north riverside are situated in the Sierra Lacandón National Park, the only PA on the Guatemalan side, which covers a major part of this side of the TRB. Nevertheless, the soil uses of grassland also coincide with a portion of the southern PA, which means that the colonization has provoked deforestation in the park and the development of livestock activities. Gallardo-Cruz et al. (2021) evidence a forest loss of 24.9% in the Sierra Lacandón National Park between 2000 and 2018, which is lower than the Guatemalan average of 37% for PAs, although it indicates a high level.

Three PAs are located on the Mexican riverside, which include a small part of the Montes Azules Biosphere Reserve in the southwest and other two small ones, the Wildlife Protection Area Chan-Kin close to the riverside in the south and Yaxchilán, which is also an archaeological site placed midstream on the international river in an omega peninsula. The soil uses indicate that the forest areas are more extended than the polygons of these three PAs, probably owing to the existence of local rules of conservation as well as some communitarian protection programs and federal incentives. The Montes Azules Biosphere Reserve registered a forest loss of 8.4% between 2000 and 2018, the most elevated for the sub-basin, whereas Chan-Kin and Yaxchilán are the less concerned with 3.7% and 0.2% (Gallardo-Cruz et al., 2021). In this case, the river boundary marks a clear difference between the effectiveness of official conservation areas, but it may also be conditioned by the extension of the PAs that likely convert them into more vulnerable and difficult spaces to conserve when they are large, such as the Sierra Lacandón National Park.

From the land border that separates Guatemala and Tabasco to the north is the Wildlife Protection Area of Cañón del Usumacinta. The Tabasco portion of the Gran Usumacinta is almost totally deforested, including parts of the Cañón del Usumacinta that was decreed in 2008, decades after the installation of rural populations that include mainly farmers and small peasants, in close proximity to big cattle ranchers. The Cañón del Usumacinta PA registered the major forest loss of the whole Mexican Usumacinta TRB between 2000 and 2018, for a total of 21% (Gallardo-Cruz et al., 2021). Thus, this protected area is adjacent to the Guatemalan Sierra Lacandón National Park. The two are separated by the land border in a remote area only accessible by road from Mexico. Our fieldwork on the Mexican side indicates a deep process of deforestation by Mexican people that enter the park and illegally extract timber to transport it to the other side, where a small Mexican village on the border has 15 carpentry shops that immediately transform it into merchandise such as furniture:

Image 4.5 Canyon del Usumacinta PA, Mexico and Timber Smuggling from the Lacandón National Park of Guatemala
Source: Edith Kauffer, July 2018

> The most part . . . let's talk, we will be clear because I like it, so I will not hide things. The most part is dedicated to cut xate palm on the other side. Most of the people are dedicated to harvesting wood from the other side, yes, and why is that? Well, because let's assume that the landless villagers can't have the same rights as us. They can't go to my plot, cut down trees, and seed. Most of the landless villagers are the ones that do that kind of work, like the wood that you just saw over there – it is probably and most likely from Guatemala. And what else do I do? I would like us all to be loyal, and all the things would go out of here. No, you can say, "Hey, don't saw because if you saw" thus "give me for food". You understand, each one resolves [his economic needs] according to his possibilities, yes. I know it is bad, I know it is forbidden, but sometimes I say: if you want me to help you, the day you fall, don't come with me. It's your problem because when the soldiers come, the minimum is 20–30 years in jail for sawing wood. . . . Mostly, I think it is about 70% of that side [that saw timber], and the rest are beans, corn, and livestock.
>
> (Fieldwork, interview with local authority, July 5, 2018)

The Ramsar site of La Libertad in the municipality of Emiliano is also a federated state PA, the second major forest loss in the Usumacinta TRB (17.5%) between 2000 and 2018, compared to an average of 8% for PAs.

Both borders of the Gran Usumacinta define differences in terms of forest coverage/deforested areas dynamics that are related to situations of access and remoteness, the presence of human settlements, and historical processes. However, TRBs facilitate not only rivers crossing from one state to another, or rivers marking the border, but also transboundary dynamics. In this case, living in a TRB generates transboundary clandestine logging from Mexico that is facilitated by contiguity. This location is also characterized by lack of opportunities for landless peasants, poverty, and a shortage of jobs and commercialization networks for selling agricultural products owing to its remoteness from political centers. As analyzed in chapters 3 and 5, these transboundary dynamics of timber logging, clandestine fauna extraction, and hunting are favored by the availability of natural resources in tropical forests and by the presence of the border because the national authorities cannot cross the dividing line and follow the infringers to the neighboring country.

The Santo Domingo TRB is characterized by a small number of protected areas in Mexico, the extension of which is reduced: a small part of the National Park of Montebello is situated in the southwest, and two Voluntarily Dedicated to Conservation Areas (ADVC, in Spanish) of Las Nubes (La Caverna) and La Serranía are located downstream. Montebello offers numerous lakes and lagoons in a pine-oak and cypress woodlands environment that has been subject to wildfires (as in 1998) and is characterized by the pollution of the lowlands lagoons, although those located in the sub-basin are highlands water bodies without pollution issues. Las Nubes is a well-known ecotourist area, owing to the presence of beautiful waterfalls. Part of the characteristics of conservation is the topography of the main Mexican watercourse. La Serranía was created in 2011, and it is also a place dedicated to ecotourism. Although small compared with the extension of the sub-basins, these three PAs were strategic when, in 2018, the Secretary of Environment and Natural Resources refused to build a hydropower plant in Santo Domingo, partly owing to the role of Las Nubes and La Serranía as the former is situated "8 km from the Montes Azules Biosphere Reserve, forming an essential part of the biological corridor between the Cuchumatanes Highlands and the Lacandon Forest, contributing to the connectivity of the landscape and genetic flow with the mobility of flora and fauna elements (SEMARNAT, 2018, 19). The importance of the Mexican Voluntary Conservation Areas is key when we take into consideration that the loss of forest in this modality of PAs was only 3.1% between 2000 and 2018, compared to 8% for all PAs (Gallardo-Cruz et al., 2021) (Image 4.6).

The Guatemalan side does not present any PA in the Santo Domingo River basin. Nevertheless, it is better conserved than the Mexican one, probably owing to the difficult mountain topography and especially the lack of livestock. Additionally, the Guatemalan NGO Fundaeco has financed 11 protection projects, among which two are part of the Santo Domingo TRB. Another difference is the primary forest extension in Guatemala versus the secondary vegetation of forest on the Mexican side, where the upstream part is more

Image 4.6 Waterfall at Las Nubes, Mexican Voluntary Conservation Area
Source: Edith Kauffer, March 2022

dedicated to agriculture and livestock, and the downstream section corresponds to the meandering part of the river to self-consumption farming and, more recently, to tourism. It is also noticeable that the Mexican portion is better accessible than the Guatemalan part, which lacks pavemented roads. For that reason, bordering communities from Guatemala cross the border to access Mexican markets and, in some cases, receive medical assistance and other public services, such as the road to commute to other bordering localities.

Forest continuity is not a transboundary characteristic in both TRB as they are above all human and social spaces and colonization areas depending on national and local dynamics. In contrast, the land and river borders show considerable differences in forest presence and deforestation dynamics in each TRB. Protected areas also play a contradictory role as their effectiveness depends on various local and national factors, as well as their location in relation to the border, their past and contemporary colonization processes, their date of establishment, and the presence of actors that depend on deforestation for their living or for conducting business. As the Guatemalan Santo Domingo TRB shows, the apparent lack of official PAs is not an obstacle to conservation. This specific local situation is reminiscent of the situation of the Guatemalan portion of the whole Usumacinta TRB, which evidences a major forest loss inside PAs (37%), compared to outside them (29%) (Gallardo-Cruz et al., 2021).

Viewed from a cross-border perspective, the forest components of these waterlands are deeply shaped by the existing international borders, and the dividing lines impose themselves on the transboundary nature of the river basins and shape a political division that not only reflects on waters but also on forests – their extension and state of conservation. Recent data clearly indicate a deep contrast regarding forest loss in Guatemala and Mexico for the Usumacinta River TRB, with major losses in Guatemala, but it also evidences a disparity about the role of PAs in each country. The major forest losses in Guatemala are located inside PAs, whereas in Mexico, they are situated outside their polygons. In this case, the border marks an important difference. Nevertheless, the two analyzed sub-basins are interesting as high-level forest losses in some PAs in both countries, but concentrated in one of them, with medium- and low-level forest losses in the Mexican case. Moreover, in both cases, PAs are not the only components of forest coverage; the forest does not only depend on the existence of governmental PAs but on the entanglements of diverse factors, some depending on topography and access, and others depending on social and political history and processes. These include the existence of other forms of local organizations for forests in the Santo Domingo sub-basin around NGOs projects or in the voluntary conservation area at a local scale in Mexico, with a low 3.1% of forest loss. Additionally, the data highlight the existence of municipal PAs that register 0% of forest loss in Mexico, as well as the high number of private areas in Guatemala, which evidence a forest loss of 15.6%, which is a lower decrease compared to the total of 37% of protected forest and 29% of unprotected forests in the whole Usumacinta TRB in that country (Gallardo-Cruz et al., 2021).

Waterless Transboundary Rivers Basins? Local Access to Tap Water and Sanitation

At the end of a workshop with local authorities in an Indigenous Tsotsil village located in a 1980s colonization area 15 km from the border with Guatemala, three men were waiting for us – two women researchers – as they wished to speak with us. They were coming from Guatemala and wanted us to visit their village because the government was building a hydropower plant without consulting their Indigenous community, and they needed help. We explained to them that we were researchers and that our work had academic purposes, but we promised to come back with members of an NGO that belonged to the Mexican Movement of People Affected by Dams and in Defense of Rivers (MAPDER in Spanish). Three weeks later, we arrived at the closest Mexican village located on the border with a colleague from the NGO. This was the unique way to reach the village located on the other side of the border. A few minutes later, the Guatemalan community authorities arrived and guided us toward the border across a dirt track, only accessible on foot, that descended to the banks of the Santo Domingo River, just where it ends, marking the boundary, and meandering into the Guatemalan territory. On our way, we passed many Guatemalans who greeted us kindly. The nearest accessible regional market was on the Mexican side, where the public services are closer than in Guatemala. To access the Guatemalan village, we first crossed the border breach, a 10-meter-wide deforested line with five meters in each country and a boundary stone – as part of a series of monuments marking the border – and two big sign boards with the states' names and colors (green for Mexico and blue for Guatemala). Once in Guatemala, we walked across the river on a hammock bridge that was recently rebuilt by the Guatemalan inhabitants after the floods had destroyed the former one. A four-wheel drive Toyota, the only vehicles that can travel over this rugged terrain, was waiting to take us to the village, which was located high in the mountains and had to be reached by an extremely steep path – a clear indication of the neglect of this village by the national government and of the local population's living conditions (Fieldwork, October 2, 2015) (Image 4.7).

Tap water and sanitation in transboundary TRBs with plenty of water that are in different points in relation to the border depend on the availability of water resources, the geographical configuration of each place, and, above all, the existence or shortage of public services. The above extract of the fieldwork diary suggests that the political border shapes an important difference in access

Image 4.7 A Hanging Bridge Built by the Guatemalan Population to Cross to Mexico in the Usumacinta TRB
Source: Edith Kauffer, December 2022

to public services. Drinking water does not exist in the houses of the Maya Forest waterlands, because tap water is not potable when it is available. This section deals with tap water and sanitation in both sub-basins to examine how waterlands might convert into waterscarce living spaces.

Tap Water in Gran Usumacinta TRB: Tap Without Water and Water Without Tap

Regarding domestic water, the Gran Usumacinta basin only offers complete data between 2000 and 2020 for Mexico, which evidences a reversal of the trend, indicating a worsening tap water situation. This could be explained by the population growth, inefficient and inadequate services, and correct update of data during the last census. It is noticeable that between 2000 and 2020 in Mexico, the data related to water for human consumption have changed from "drinking water" – albeit never adequate and potable to be consumed as such – to new categories. Owing to these divergences, I decided to differentiate two possible realities: the existence or inexistence of tap water in the home. For the year 2010, I also chose to present data on the presence of water in the garden because, in the Guatemalan case, data from 2018 detailed the location of water or type of access.

In fact, during our fieldwork, we often come across situations where tap water is not available, even though the statistics indicate that the inhabitants have water in their house. They may have a tap but no water flowing from it. Often, damages to the rural water system are not repaired, and tap water converts into a waterless tap. In other cases, when public services are totally inexistent, the whole village goes to the river or to a public well.

Although access to tap water at home in Mexico is low (this service is not available to 10.2%–47.6% of the population), the conditions in the Guatemalan part of the TRB are even worse, according to Table 4.1: between 62.1% and 87.97% of people live with a tap water shortage in their house. A critical situation

Table 4.1 Tap Water in the Gran Usumacinta TRB

Municipality	*% Tap water in houses 2000*	*% Tap water in houses 2010*	*% Tap water in garden 2010*	*% Without tap water in house 2018 (G) & 2020 (Mx)*
Balancán (Mx)	80.06	84.67	14.95	47.6
Emiliano Zapata (Mx)	87.19	80.09	19.74	19.9
Tenosique (Mx)	68.93	79.34	20.36	38.6
Ocosingo (Mx)	81.45	83.33	16.19	14.7
Palenque (Mx)	73.49	71.51	28.05	10.2
La Libertad (G)	n.d	n.d	n.d	62.1
Las Cruces (G)	-	-	-	87.97

Mx: Mexico; G: Guatemala
Source: INEGI 2000, 2010, 2020; INE 2002, 2018.

is the fact that, as in Balancán, Tenosique, and Las Cruces, direct access to the river does not mean a better access to tap water because public services are likely not considered necessary when the natural abundance of water guarantees other types of access to water outside the house in this TRB (Image 4.8).

To illustrate the shortage of tap water at home in the Guatemalan case, Table 4.1 presents data to compare with the whole department of Petén and understand the origin of water for human consumption. People living in Las Cruces and La Libertad municipalities have less access to tap water inside the house compared to the rest of the department, which evidences the lack of services and infrastructure due to the remoteness and general shortage of access to this area. In Las Cruces, tap water in the house (12.03%) ranks fourth after public or private wells (42.52%), tap water in the garden (18.27%), and public tap (14.64%). Without exception, when navigating on the Usumacinta watercourse by boat, on the Guatemalan riverside, women always wash clothes and groom children. However, this is never the case on the Mexican riverbank, which points to a deep difference between both populations. Despite the river being a source of running water for the local population, without considering the issue of water quality due to the pollution of the rivers, this is not the case in Guatemala. Nevertheless, in Las Cruces, the abundance of water explains that wells, water harvesting, natural spring, and water from the river represent 52.53% of the type of access in the whole municipality. Thus, the presence of women conducting traditional domestic activities on the riverbanks is a logical illustration of the shortage of tap water at home and in the garden (Image 4.9).

Image 4.8 Water Pumping in the Usumacinta River in Las Cruces, Guatemala
Source: Edith Kauffer, June 2022

Image 4.9 Women Washing and Bathing the Children in the Usumacinta River in Las Cruces, Guatemala
Source: Edith Kauffer, June 2022

In La Libertad, only 37.9% of inhabitants have tap water at home, which presents a better situation than Las Cruces. The majority (43.47%) have access to a tap in the garden, and some families access water from public and private wells (9.71%).

The abundance of water guarantees natural access to sources, but the shortage of infrastructure constitutes the main issue in both parts of the TRB. The conditions of access to water in the Guatemalan part are even worse, according to Table 4.1 and Table 4.2, which evidences a shortage of public services. In the Gran Usumacinta TRB, sharing waters and living on the major river of the region or in its river basin does not warrant access to adequate water for human consumption and is characterized by a substantial difference of access between the Mexican and Guatemalan riversides.

Santo Domingo TRB: "The unique dry place is the water tap"

In November 2020, two rainstorms, Eta and Iota, hit the Usumacinta River basin, especially the upstream Guatemalan part. Bulej, the most populated village of the Guatemalan Santo Domingo TRB was affected by flooding, and as the title of a newspaper indicates, "the unique dry place in Bulej is the water tap" (Fehrm, 2020). This evokes a great contradiction that characterizes the Maya Forest waterlands. Bulej, an Indigenous Chuj community, is one of the

Table 4.2 Water Access in Petén, 2018

Type of access	*Petén*	*Las Cruces*	*La Libertad*
Tap water in house	57.17%	12.03%	37.90%
Tap water outside	31.36%	18.27%	43.47%
Public tap	1.16%	14.64%	2.94%
Public or private well	3.56%	42.52%	9.71%
Rainwater harvesting	2.42%	3.12%	2.05%
River	0.15%	2.99%	0.15%
Lake	0.93%	0 %	0.66%
Natural spring	0.78%	3.90%	1.67%
Water tanker	1.95%	0.82%	0.34%
Other	0.53%	1.70%	1.13%

Source: INE 2018.

rainiest places of Santo Domingo and the whole Usumacinta TRB. During those rainstorms, the complete village was converted into water: crop fields and roads suddenly became lakes and rivers (i.e., the concrete expression of waterlands where land and waters merge). Above all, however, Bulej suddenly transformed into the clear example of the major contradiction mentioned in this chapter: plenty of water but waterless for human basic needs.

The Santo Domingo TRB only registers the complete chronology for the Mexican section and data from 2002 and 2018 for Guatemala (Table 4.3). In Mexico, the situation of access to tap water is highly contrasting and sometimes contradictory, across the decades. Regarding the access to tap water at home, Las Margaritas registers the best situation and Maravilla Tenejapa the less favorable, perhaps owing to its condition as a new municipality and past remoteness from the former capital of the municipality.

Table 4.3 Tap Water in Santo Domingo TRB

Municipality	*% Tap water 2000 & 2002*	*% Tap water inside house 2010*	*% Tap water in garden 2010*	*% Without tap water in house 2018 & 2020*
Altamirano (Mx)	23.07	0	100	26.5
La Independencia (Mx)	32.87	73.94	25.77	30.3
La Trinitaria (Mx)	64.48	35.37	50.46	34.7
Las Margaritas (Mx)	56.53	66.07	33.48	9.9
Maravilla Tenejapa (Mx)	43.74	84.79	14.39	44.3
Barillas (G)	49.96	n.d	n.d	65.73
Nentón (G)	49.05	n.d	n.d	72.13
San Mateo Ixtatán (G)	49.23	n.d	n.d	81.68

Source: INEGI 2000, 2010, 2020; INE 2002, 2018.

In Guatemala, access to tap water in 2002 and 2018 does not indicate a better situation and has huge differences with Mexico, although the latest data have shown worsening results. The situation of Bulej, located in Mateo Ixtatán, mentioned above, completely aligns with this reality of a tap water shortage in the Guatemalan upstream TRB.

As shown in Table 4.4, the lack of water services in the house in the Guatemalan part of the Santo Domingo sub-basin is more acute in the bordering municipalities than in the whole department. It is notable that the most remote part of the basin, San Mateo Ixtatán, is characterized by the lowest percentage of tap water in the house. It shares with Nentón the highest percentage of access to water consisting of tap water outside the house, a typical situation for rural communities. Barillas offers anyway more infrastructure: water harvesting is likely promoted by NGOs' projects according to fieldwork in the region, and access to water through wells is developed in the lower lands. In Nentón, the river is a source of water for housing, whereas in Barillas, springs are available. In a context of shortage of water network-supplying households, access to natural sources depends on the local context and the presence of springs, rivers, or lakes.

As in the Gran Usumacinta TRB, the Santo Domingo TRB evidences a situation of general shortage of access to tap water at home, although more acute in Guatemala than in Mexico. Living in a TRB watered by a big river, as in the Gran Usumacinta, or drenched by plentiful rainfall, as in the Santo Domingo TRB, does not warrant access to water for human consumption at home. Furthermore, the borders – fluvial or land boundary – define a better or worse access to water services, depending on the side of the border where the population is living.

Waterless TRBs, Although Full of Sewage

Close to the beginning of the upstream rapids on the main watercourse, there is a last village on the Mexican riverside that hosts a small Indigenous community

Table 4.4 Water Access in Huehuetenango, 2018

Type of access	*Huehuetenango*	*Barillas*	*Nentón*	*San Mateo Ixtatán*
Tap water in house	52.06%	34.27%	27.87%	18.32%
Tap water outside	4.21%	24.92%	52.12%	75.10%
Public tap	1.39%	5.52%	5.37%	1.17%
Public or private well	37.65%	7.59%	2.66%	1.88%
Rainwater harvesting	0.46%	15.49%	1.76%	1.31%
River	0.24%	0.83%	4.19%	1.43%
Lake	0%	0.01%	0.18%	0%
Natural spring	0.98%	10.17%	1.42%	0.53%
Water tanker	1.96%	0.06%	1.07%	0.01%
Other	1.06%	1.15%	3.37%	0.26%

Source: INE 2018

of Tseltal and Chol population (84%) with a monolinguism rate of 27% compared to only 6.8% in the whole municipality (INE, 2000). From this point, which is situated in the middle of the Gran Usumacinta TRB, it is possible to navigate upstream to the biggest sediment beach known as "El Desempeño" and, by boat, sail downstream to the archaeological site of Piedras Negras, which is only accessible from the south through the river. This Mexican village is reachable by a dirt road that becomes muddy with the rain, and the last kilometers end with a slightly hilly relief. The population lives in conditions of high marginality, as shown by the 15 localities with the most service shortages in the municipality of Palenque, Chiapas (Sedesol, 2014) and the 70 localities with higher levels of social marginalization (Sedesol, 2016). Part of the population does not speak fluent Spanish and only attends primary school for a few years. Regarding access to water networks and tap water in houses, services are nonexistent, although the inhabitants live in a place where they can use the big river. Moreover, houses have no toilets or latrines, and people go behind bushes, as observed during fieldwork (Image 4.10).

Along the whole Usumacinta TRB, fieldwork and data evidence that sanitary conditions are worse than the access to tap water. The following data only considers access to a sewer in houses as part of sanitation services – that is, the available shared data between the two countries. As for tap water, the complete chronology only exists in Mexico, and the Guatemalan data are available for 2002 and 2018.

Image 4.10 Latrines in Guatemalan Colors at El Ceibo Border Checkpoint
Source: Edith Kauffer, July 2018

Although access to the sewer in the Mexican section of the Gran Usumacinta was generally better in 2020 than in 2000, the case of Ocosingo seems to be different, probably in some villages that are not included in the TRB. Nevertheless, the Guatemalan situation points to sewer conditions that are almost nonexistent in the Gran Usumacinta TRB (Table 4.5).

In the Santo Domingo TRB, the Mexican part is characterized by a growing number of sewers and a declining percentage of villages without sewers, excepting Las Margaritas, which registered a growing shortage in 2020 (Table 4.6). In Guatemala, the 2000 data evidence the absence of sewers (more than 99%), and in 2020, the shortage was considerably lower (about 70%–72%). However, the difference between Mexico and Guatemala regarding sewers is enormous.

Both tap water in houses and the shortage of sewers in the studied transboundary sub-basins evidence a huge gap between Mexico and Guatemala in both living spaces. Although the access to tap water is not good in Mexico, the

Table 4.5 Sewer in Gran Usumacinta TRB

Municipality	*% sewer in houses 2000*	*% sewer in houses 2010*	*% without sewer 2010*	*% without sewer 2018 & 2020*
Balancán (Mx)	91.53	96.55	2.85	13.5
Emiliano Zapata (Mx)	91.46	96.90	2.61	1.3
Tenosique (Mx)	84.07	94.72	4.47	3.9
Ocosingo (Mx)	13.68	41.77	57.89	64.4
Palenque (Mx)	17.97	67.85	30.75	5.1
La Libertad (G)	n.d	n.d	n.d	93.83
Las Cruces (G)	-	-	-	96.95

Source: INEGI 2000, 2010, 2020; INE 2002, 2018.
Mx: Mexico; G: Guatemala

Table 4.6 Sewer in Santo Domingo TRB

Municipality	*% with sewer 2002*	*% with sewer 2010*	*% without sewer 2010*	*% without sewer 2018 & 2020*
Altamirano (Mx)	0	1.97	98.03	26.6
La Independencia (Mx)	26.05	38.06	61.94	26.8
La Trinitaria (Mx)	18.04	33.77	66.23	23.7
Las Margaritas (Mx)	25.53	45.53	54.47	61.4
Maravilla Tenejapa (Mx)	1.88	52.42	47.58	45.5
Barillas (G)	0.24	n.d	n.d.	72.40
Nentón (G)	0.47	n.d	n.d.	70.90
San Mateo Ixtatán (G)	0.46	n.d	n.d.	70.02

Mx: Mexico; G: Guatemala
Source: INEGI 2000, 2010, 2020; INE 2002, 2018.

Guatemalan situation is worse. Further, regarding sewers, the data show a larger difference between both sections of the river basin. This means that the border determines differentiated access to public tap water and wastewater services that depend on national policies. Consequently, although people share waters, forest, wetlands, and a common universe of waterlands, the existence of political borders that depend on sovereignty, state's interest, and territorial control divides the TRB; it segregates and disconnects both realities, which are deeply entangled, and creates inequalities between neighbors that are only separated by porous land and fluvial borders.

With the better access to public water and sanitation services on the Mexican side in both TRBs and the general conditions of access through pavemented roads, local markets have favored transboundary exchanges of services, such as the delivery of water from Guatemala by gravity instead of electricity, health services, or the use of pavemented roads from Mexico (Kauffer Michel, 2013; Kauffer, 2018).

As evidenced for TRBs in general, the political border is a component that goes against the tide and opposes the idea of sharing natural resources, such as water or forests. International borders divide TRBs and inhibit cooperation around the environment as the existence of TRBs is considered a threat to states' sovereignty. The evidenced situations regarding the differences in access to water and sanitation services between Mexico and Guatemala create inequities between fragmented neighbors by a porous line or by a local river that connects people more than separating them.

Conclusions

The transboundary river basins between Mexico and Guatemala are not officially recognized by both states, which impedes cooperation and the minimal consideration that consists in creating a joint mapping. In this context, water sharing and transboundary water issues only include local stakeholders and some NGOs' attempts at projects (Kauffer, 2018). The sole mention of the idea of TRBs is banished in certain political circles, and research about them is discredited and sometimes circumscribed. Initiatives have been systematically boycotted by Guatemala and generally ignored by Mexico, as evidenced by two decades of fieldwork at the water policy scale and on transboundary issues.

TRBs are not only concerned with waters but also with environmental conditions and the Maya Forest waterlands, specifically with the presence and conditions of the tropical forest. Forest and water are entangled as empirical realities, but also analytically, through the concept of waterlands proposed in this book.

In such a way, TRBs exist as geographical realities and are pertinent as a remote dream of joint management at the international scale from a shared International Water Resources Management perspective that includes waters, forests, and their entanglements. However, they are also conceptually substantial to understand and demonstrate a major contradiction of the Maya Forest waterlands – that is, a world full of waters that coexists with water scarcity for the domestic uses of the local population.

Above and beyond water confluences and rivers, the analysis of TRBs presented in this chapter also offers an understanding of the characteristics of the local environment in borderlands contexts. Moreover, this chapter introduces the role of the political border for local populations in two transboundary sub-basins and reveals that the border not only defines the relations with waters but also with forests. Additionally, it marks deep contrasting realities between Mexico and Guatemala regarding the access to water and sanitation services, the state of forest coverage, and the role of PAs. Consequently, these findings suggest that TRBs are more divided by the border than united by the flowing waters.

Nevertheless, as part of the Maya Forest waterlands, TRBs constitute spaces of continuity and offer possibilities of exchanges on shared resources. This occurs at the local scale, although conflicts sometimes arise. Waters and fauna cross borders, but other natural resources, such as wood, animals, and sediments, are transported by people crossing the borders. Forest-waterlands peoples share not only waters and forests but also the inequalities that characterize the populations of borderlands and areas of colonization in different parts of the world and in Mesoamerica. With or without the line, they share conditions of marginality related with the remoteness of their location and the contradictory lack of presence of the states that supposedly and contradictorily claim to defend their borders and have historically promoted the settlements to mark the limits of their national territories.

The Maya Forest waterlands, as worlds with plenty of water, are also waterless living spaces for local inhabitants, creating waterwaste issues for downstream neighbors, both close and far away. How can one live on the banks of a great river without accessing tap water at home? How can one send wastewater to the neighbor who is located downstream, owing to the shortage of treatment plants and sewage? How can one combine remoteness, state failure with the rural population, and the ability to look for local and creative solutions?

The borderlands context that defines waterlands and the political borders that also determine the existence of TRBs converged in this chapter to explain how a waterful world converts into a waterless or waterscarce living environment for riverine and local populations. These analytical perspectives complement each other to highlight common contradictions and huge contrasts. Living in borderlands does not indeed imply the same living conditions as it entirely depends on which side of the border or of the international river people are settled. Therefore, the Maya Forest waterlands, analyzed from a TRBs perspective based on empirical results that combine qualitative information and quantitative data, reveal that despite the attempts to go beyond boundaries, considering shared resources (TRBs) and the mobile convergence of water and land (waterlands), borders as markers of states are ever present.

Notes

1 The difference between the Usumacinta TRB and Gran Usumacinta transboundary sub-basin will be explained in the following pages as part of the river basins' description.

2 The Gran Usumacinta transboundary sub-basin is one of the seventeen sub-basins of the complete Usumacinta River, the major TRB. Owing to the nested feature of river basins, when considered as part of the major TRB (the Usumacinta River basin), both the Gran Usumacinta and the Santo Domingo sub-basins may be designated as transboundary sub-basins. Nevertheless, when analyzed separately, both sub-basins are also TRBs (i.e, river basins that transcend a border). Consequently, they are considered both transboundary sub-basins and TRBs depending on the perspective.

References

Aguilar, G. & Iza, A. O. (2009). *Gobernanza de aguas compartidas: aspectos jurídicos e institucionales* (No. 58). IUCN.

Allouche, J. (2010). The multi-level governance of water and state-building. In K. Wegerich & J. Warner (Eds), *The politics of water: A survey* (pp. 45–67). Routledge.

Camargo, A. (2022). Land born of water: Property, stasis, and motion in the floodplains of northern Colombia. *Geoforum*, 131, 223–231.

Caponera, D. A. (1993). Legal aspects of transboundary river basins in the Middle East: The Al Asi (Orontes), the Jordan and the Nile. *Natural Resources Journal*, 33, 629–663.

Cascão, A. E., & Zeitoun, M. (2010). Power, hegemony and critical hydropolitics. In E. Earle Anton, A. Jägerskog & J. Öjendal (Eds), *Transboundary water management: Principles and practice* (pp. 27–42). Earthscan.

Gallardo-Cruz, J. A., Peralta-Carreta, C., Solórzano, J. V., Fernández-Montes de Oca, A. I., Nava, L. F., Kauffer, E., & Carabias, J. (2021). Deforestation and trends of change in protected areas of the Usumacinta River basin (2000–2018), Mexico and Guatemala. *Regional Environmental Change*, 21, 1–15.

García García, A., & Kauffer Michel, E. F. (2011). Las cuencas compartidas entre México, Guatemala y Belice: un acercamiento a su delimitación y problemática general. *Frontera Norte*, 23(45), 131–161.

Ghiotti, S. (2006). Les territoires de l'eau et la décentralisation. La gouvernance de bassin versant ou les limites d'une évidence. *Développement durable et territoires. Économie, géographie, politique, droit, sociologie*. doi:10.4000/developpementdurable.1742.

Ghiotti, S. (2014). Le bassin transfrontalier. *Regions and Cohesion*, 4(2), 10–29.

INBO-GWP (International Network of Basin Organisations – Global Water Partnership) (2009). *A Hanbook for integrated water management in river basins*. http://www.gwp.org/globalassets/global/toolbox/references/a-handbook-for-integrated-water-resources-management-in-basins-inbo-gwp-2009-english.pdf.

INE (Instituto Nacional de Estadística). (2002). *XI Censo Nacional de Pobñlación y VI de Vivienda*, https://www.ine.gob.gt/censo-2002/.

INE (Instituto Nacional de Estadística). (2018). *XII Censo Nacional de Población y VII de Vivienda*, https://censo2018.ine.gob.gt

INEGI (Instituto Nacional de Estadística y Geografía). (2000). *XII Censo de Población y Vivienda 2000*. https://www.inegi.org.mx/programas/ccpv/2000/

INEGI (Instituto Nacional de Estadística y Geografía). (2010). *Censo de Población y Vivienda 2010*. https://www.inegi.org.mx/programas/ccpv/2010/

INEGI (Instituto Nacional de Estadística y Geografía). (2020). *Censo de Población y Vivienda 2020*. https://www.inegi.org.mx/programas/ccpv/2020/

Kauffer, E. (2018). Un análisis de la gobernanza multinivel en las aguas transfronterizas de México con Guatemala y Belice. *Frontera Norte*, 30(60), 31–56.

Kauffer, E. F. (2019). From the abundance of waters to the scarcity of studies: Contemplating hydropolitics in Mexico-Guatemala and Mexico-Belize borders. In H. R. Guerrero García Rojas (Ed.), *Water policy in Mexico: Economic, institutional and environmental considerations*, (pp. 237–262). Springer.

Kauffer, E. (2020). Entre una GIRH inconsistente e inseguridades hídricas multiformes: la cuenca transfronteriza del río Hondo en la política mexicana/Between an erratic IWRM and multifaceted water insecurities: The transboundary Hondo River Basin in Mexican policy. *Tecnología y Ciencias del Agua*, 11(6), 127–176.

Kauffer, E. (2024). Beyond conflict and cooperation in Central American transboundary river basins: Crossroads of actors and borders. In L. P. B. da Silva, W. C. Ribeiro & I. B. Espíndola (Eds.), *New perspectives on transboundary water governance. Interdisciplinary approaches and global case studies* (pp. 61–80). Routledge. doi:10.4324/9781003333678-7.

Kauffer Michel, E. F. (2011). De la abundancia del agua a la escasez de estudios: retos y perspectivas de las hidropolíticas en la frontera México-Guatemala-Belice. In E. F. Kauffer Michel (Ed.), *Entre manantiales y ríos desatados: paradojas de las hidropolíticas fronterizas (México-Guatemala)* (pp. 9–31). Centro de Investigaciones y Estudios Superiores en Antropología Social – El Colegio de Michoacán.

Kauffer Michel, E. F. (2013). Represas en la cuenca transfronteriza del río Usumacinta: Un conflicto crónico. In E. F. Kauffer Michel (Ed.), *Cuencas en Tabasco: Una visión a contracorriente* (pp. 101–132). CIESAS.

Kauffer Michel, E. F. (2018) Introducción: cuencas transfronterizas en la encrucijada de los conflictos y la cooperación: Lo político más allá del nacionalismo metodológico. In E. F. Kauffer Michel (Ed.), *Cuencas transfronterizas: La apertura de la presa del nacionalismo metodológico* (pp. 15–44). CIESAS.

Kauffer, E., & Maganda, C. (2022). The adoption of global water norms in Central America: What separates normative coherence from normative hegemony? *Development Policy Review*, 40, e12626.

Lasserre, F. & Boutet, A. (2002). Le droit international réglera-t-il les litiges du partage de l'eau? Le bassin du Nil et quelques autres cas (Note). *Études Internationales*, 33(3), 497–514.

Leyva, X., & Burguete, A. (Eds.) (2007). *La remunicipalización en Chiapas: La política y lo político en tiempos de contrainsurgencia*. Centro de Investigaciones y Estudios Superiores en Antropología Social.

Leyva Solano, X. & Rodríguez Castillo, L. (2007). Espacios disputados y redes clientelares en la formación del municipio de Maravilla Tenejapa (región Selva Fronteriza). In A. Burguete Cal y Mayor & X. Leyva Solano (Eds.), *La remunicipalización en Chiapas: La política y lo político en tiempos de contrainsurgencia*. (pp. 179–223). Centro de Investigaciones y Estudios Superiores en Antropología Social.

Milich, L., & Varady, R. G. (1999). Openness, sustainability, and public participation: New designs for transboundary river basin institutions. *The Journal of Environment & Development*, 8(3), 258–306.

Mirumachi, N. (2015). *Transboundary water politics in the developing world*. Routledge.

Mirumachi, N., & Chan, K. (2014). Anthropocentric hydro politics? Key developments in the analysis of international transboundary water politics and some suggestions for moving forward. *Aquatic Procedia*, 2, 9–15.

Olvera Alarcón, D. N., Kauffer Michel, E. F., Inge Schmook, B., & Huicochea Gómez, L. (2011). Factores de conflicto en la cooperación por el agua en cuencas compartidas: caso Río Hondo (México-Guatemala-Belice). *Estudios Fronterizos*, 12(24), 103–134.

Sadoff, C. W., & Grey, D. (2005). Cooperation on international rivers: A continuum for securing and sharing benefits. *Water International*, 30(4), 420–427.

Sedesol (Secretaría de Desarrollo Social). (2014). *Informe anual sobre la situación de pobreza y rezago social*. https://www.gob.mx/cms/uploads/attachment/file/32855/Chiapas_065.pdf.

Sedesol (Secretaría de Desarrollo Social). (2016). *Informe anual sobre la situación de pobreza y rezago social*. http://diariooficial.gob.mx/SEDESOL/2016/Chiapas_065.pdf.

Semarnat (Secretaría de Medio Ambiente y Recursos Naturales). (2018). *Rechaza Semarnat Presa Hidroeléctrica Santo Domingo*, Chiapas, https://www.gob.mx/semarnat/prensa/rechaza-semarnat-presa-hidroelectrica-santo-domingo-en-chiapas.

Sneddon, C., & Fox, C. (2006). Rethinking transboundary waters: A critical hydropolitics of the Mekong basin. *Political Geography*, 25(2), 181–202.

Teasley, R. L., & McKinney, D. C. (2011). Calculating the benefits of transboundary river basin cooperation: Syr Darya Basin. *Journal of Water Resources Planning and Management*, 137(6), 481–490.

UNEP-DHI (United Nations Environment Programme). (2016). *Transboundary river basins: status and trends*. Vol 3. Nairobi: United Nations Environment Programme (UNEP). www.geftwap.org/publications/river-basins-spm.

Waterbury, J. (1997). Between unilateralism and comprehensive accords: Modest steps toward cooperation in international river basins. *International Journal of Water Resources Development*, 13(3), 279–290.

Zeitoun, M., Goulden, M., & Tickner, D. (2013). Current and future challenges facing transboundary river basin management. *Climate Change*, 4(5), 331–349.

5 Political Trails in the Maya Forest

Go-Betweens, Curating, and Places-in-Knots in Three Biological Stations

Hanna Laako

Introduction

> In this vast and all but uninhabited, heavily forested region the archaeologist perforce must follow the ways blazed by chicleros or "chewing gum hunters"; their trails must become his trails, their camps and water holes his camping places, their mule trains his only means of transportation from site to site. In the explorations of Petén, the archeologist owes more to the chicle business and its far-reaching bush ramifications than to any other single agency.
>
> (Morley, 1938, viii)

Despite its political borders and distinct regional processes that shape the Maya Forest waterlands in time and place, some histories are shared within the forest shadows and extend their trails and layers into today's conservation and archaeological work. This shared regional history is intimately connected to Mayan ruins in the form of archaeological work and is currently materialized in many protected areas and biological stations. One defining characteristic of this contemporary history is that of *chiclería* – the production of chicle, a natural chewing gum that integrated the Maya Forest into the global exportation market from about the 1890s to the 1980s and served as the economic basis of the region, which included new settlements. Schwartz (1990) calls this the social history of the white gold (*oro blanco*) that formed a forest society in Petén, Guatemala, but which importantly extends to the Belizean and Mexican forest regions – the Maya Forest.

What is more, the *chiclero* tappers and hunters, in their search for the sapodilla trees (*Manilkara zapota*) to extract the nontimber product, form the roots of two other contemporary key phenomena of the Maya Forest. First, their trails and camps served and keep serving as the routes and stations of the archaeologists and conservationists working in the Maya Forest space. Second, beyond the factual forest paths, it is their accumulated ecological knowledge of their forest waterlands that has helped in the formation of the biological and archaeological stations that now feed into our scientific knowledge of the hotspot and its diversity. These two aspects are well illustrated in the above citation by the archaeologist Sylvanus Morley (1883–1948), who explored the ruins mainly guided by *chicleros*, or by using the *chiclero* trails – in other words, the wayfinding *chiclero* knowledge related to the Maya Forest waterlands.

DOI: 10.4324/9781003429050-6

While the *chiclero* epoch has now ceased, the nearly hundred years of the industry formed the history of a political forest based on its trails, routes, and camps/stations, which are located nearby waterholes (*aguadas*) and are now the *loci* of the new forest actors. These new forest actors include, among others, the previously mentioned archaeologists and conservationists who build on that shared history and knowledge but also new categories, such as the forest concessionaries and Indigenous communities established in what were previously the *chicle* camp places. The trails have been, of course, also used by oil explorers, militaries, paramilitaries, and rebels of all kinds. This chapter cannot adequately address them all; it is rather focused on the formation of the Maya Forest trails linked to conservation and archaeology. These latter actors and their shared *chiclero* places now deal with other complex political phenomena that characterize the forest society today, such as geopolitical tensions between countries, Indigenous, and peasant land-rights struggles, ecotourists, and drug-cartel activity. Thus, they are also contemporary political forests that question and address different borders, while forming, as Saxer (2022) puts it, places-in-knots, that is, tightly enlaced localities. The contemporary conservation paths are like windows that connect to Maya Forest's often-neglected transboundary socioenvironmental histories, as well as contemporary dynamics that position conservationists as mediators, witnesses, and actors within new cross-regional processes.

Peluso and Vandergeest (2011) coined the term *political forests* to understand how forests became part of national projects, interventions, and controls (particularly in Southeast Asia) when mapped and defined as such. These forests were defined according to the Food and Agriculture Organization (FAO) and were separated from other land uses, such as agricultural zones. Political forests then became tools to control and intervene in national woodlands. Political forests, for Peluso and Vandergeest (2011), are about the making of national territories subject to legislation and mapping, while extending control over illicit or rebel activities – or what could be defined as "illicit" or "rebel" because of the making of national territories. Political forests relate to forest waterlands by enclosing and separating forests from waters and other land uses, by placing them firmly within national boundaries, and by recognizing them as such units: As national forests, often subject to forestry practices.

However, this chapter seeks to show that such definition of a political forest, tied solely to marking national territories by means of mapping and legislation, may also extract the *political* out of the forest, and collocates these territories as abstract, lifeless, and bordered material-spatialities. It eliminates their political and social histories and geographies, which, as I argue, make the political forests, such as the *chiclería* that extend to contemporary archaeological sites and biological stations that form the Maya Forest. The history of the Maya Forest waterlands suggests the construction of a political forest that precisely addresses transboundary tensions and cross-regional histories and challenges, even geopolitics. The Maya Forest as a political forest also comprises its complex socioenvironmental histories that the states occasionally map out or cause by cartographic ambitions. The contemporary biological and archaeological stations and their work allow to carve out the multiple layers, trails, and camps of a political forest called the Maya Forest, and as an ecoborderland that mediates between other bordered lands. Political forests are political precisely by addressing the multilayered, entangled, and contested geographies and histories of forests, which connect to different scales and actors. As suggested in our theoretical frame (Chapter 1), the Maya Forest waterlands are political borderlands that include such state interventions, mapping, and legislation as coined by Peluso and Vandergeest (2011); however, they are also *political* for questioning these, for entailing a set of actors that are not only governmental, but that relate to livelihoods and go-betweens, and for entailing the contested perceptions over the place and politicized lands that extend beyond national borders or – even more – draw their own ones.

To this end, this chapter explores a different angle to political forests, with a focus on historical layers, trails, and places-in-knots that are simultaneously shared, entangled, contested, and (occasionally) silenced. They are also places for new agencies and the revival of historical ones. For this purpose, the chapter draws from our theoretical frame of political borderlands and particularly Metcalf's (2005) and Prado's (2012) middle grounds and *go-betweens*, which refer to hybrid spaces that lack clear division and are characterized by fluid edges and encounters where new alliances are created. Go-betweens are agents that connect different societies physically, socially, and culturally. This is precisely what the *chicleros* did. Their agency extended the historical layers through today's biological stations of the Maya Forest eco-borderlands and biodiversity hotspot. Building on Metcalf (2005) and Prado (2012, p. 323), go-betweens master certain values, practices, and status in more than one "world", which makes them crucial agents of colonization and the integration of new geographical and social landscapes. Go-betweens do not necessarily pertain to a specific ethnic origin or gender, but they are rather defined by their roles as cultural mediators that may be conceptualized in three categories (Metcalf, 2005): Physical (e.g., sea captains, sailors, passengers, slaves), transactional (translators, traders, cultural brokers), and representational (chroniclers, mapmakers, writers, orators, artists). In fact, Prado (2012, p. 323) points out that many imperial or colonial agents, such as bureaucrats, mapmakers, and military men, relied on their local informants and networks to navigate these societies and places. In many ways, the *chicleros* – as do the conservationists today – can be perceived as go-betweens.

While the bureaucratic state-apparatus, with its colonial mapmakers charting national territories may sustain Peluso and Vandergeest's (2011) political forests, this chapter explores the rather entangled, contested, mediated, historically layered, and geographically transboundary political forest, which also builds on Ingold's (2000) wayfinding, Saxer's (2022) places-in-knots, and Tuck and McKenzie's (2015) places. All three previously mentioned are conceptually related in their understanding of the place as relational land, characterized by connectivity, storytelling, and movements opposed to abstract territorial, cartographic representations in maps.

For Ingold (2000, pp. 219–242), wayfinding and mapping are nearly synonyms. Both inevitably involve storytelling about one's experience of moving between places. Places, for Ingold, are about histories rather than mere locations. These places are connected by movements – wayfinding, mapping – that when taken together make up a *region*. This kind of wayfinding and mapping is based on the knowledge-in-the-making of condensed histories, which is different from mapmaking, according to Ingold (2000, p. 220). Such knowledge is fundamentally *un*-map-like because mapmaking eradicates movement and stories, thus divorcing knowledge from the bodily experience of the need to travel. Ingold (2000, p. 234) calls this kind of mapmaking, as illustrated in Peluso and Vandergeest's (2011) political forest, a *cartographic illusion*: A theatrical stage from which all the actors have mysteriously disappeared, which is deserted and devoid of life, conveying an illusion of well-defined borders and control. In cartographic illusion, no one is there, and nothing is happening. Instead, in mapping and wayfinding, the world has no frames, and it is not limited to the surface.

The way in which Ingold (2000) refers to wayfinding and mapping as a movement between places, building on the experiences and stories that form a region, is similar to that of Tuck and McKenzie (2015). For them, a place is different from a space in that places are defined as sites of presence, futurity, power, and knowing. They critically point out that, while initially the ecologists integrate the element that is most often missing from mapmaking – that is, the ecosystems – they end up separating it from the place, thus reducing complex ecologies to mere metaphor, emptying them of their connections to land and the environment. Despite detecting incommensurability in Environmental and Indigenous Studies, they argue that both are necessarily intertwined since they have a particular "material turn" in their analysis: Integrating the lived place in the mapmaking, instead of treating it as an empty surface. Indeed, as this chapter shows, biological stations are physically located in places. For Tuck and McKenzie (2015), a spatial vocabulary has a colonial tone to it (e.g., *line, center, outside*), whereas place is historical and context specific. Yet, Tuck and McKenzie (2015) maintain that places do not always have names or are not always justly named, and they do not always appear on maps or with agreed-upon boundaries. In this sense, places escape Peluso and Vandergeest's (2011) political forest; however, their wayfinding may discover other types of political forests.

Finally, Saxer (2022) builds on wayfinding places by developing the notion of *places-in-knots* in the case of Himalayas and beyond. His finding is that seemingly remote (often perceived as isolated, frozen in time, marginal, out-of-the-place) Himalayan villages are actually globally connected "knots", thus relational places that consider themselves as the opposite of remote. Instead, for Saxer (2022), remoteness is something actively made and unmade, embedded in larger political and economic agendas rather than a space that is a prehistorically fixed leftover area. In his view (2022),

> once an area surfaces on the radars of states, development agencies, or mining companies as 'remote', it does so with certain ambitions that already carry the seed of a future relation involving the selective unmaking of remoteness for a specific purpose
>
> (p. 12)

As Torras (2019) notes, remoteness and marginality are often tools that allow both control and dissolution of certain structures, where subalternity, which is often perceived as the equivalent of an empty space, is an active part of capitalist reproduction. In other words, supposed margins or remote places are, in fact, transitory spaces. One of these recent radars of unmaking remoteness is, for Saxer, *curating* by development agencies and conservation organizations that seek to heal the fragile ecosystems and peoples suffering from so-called underdevelopment, as well as to safeguard cultural and natural heritage. Curating, while facilitating certain opportunities and connections, may also foster remoteness for preservation and thus ignore existing places-in-knots, which for Saxer (2022) are not just harmonious or cohesive, but entangling. Yet, they are also persistent and can witness revivals after long phases of decline, particularly in places lacking dense networks of roads and railways. Thus, political forests are about the relations and regions born in the places-in-knots that are central for mobilities, trade, and connectivity even in seemingly remote places, such as the Maya Forest Waterlands.

This chapter first discusses the *chiclero* history shared by the Maya Forest. The section shows how the *chicleros* are key actors, or go-betweens, in enabling the creation of the Maya Forest in space and knowledge by archaeologists and conservationists later on. Second, the chapter introduces three cases, which are all biological stations run by conservationists in the Maya Forest borderlands and are often also spaces for archaeological work. The entangled history of the biological stations is narrated with particular attention to the place and places-in-knots, and thus, conservationists also become new kinds of go-betweens, while intending to curate fragile ecosystems in the Maya Forest Waterlands.

The chapter is based on fieldwork in the three biological stations, participant observation, and interviews carried out between the years 2018 and 2023. Varied first and secondary materials have been used for the *chiclero* history, such as Morley's diaries and published work, visits to *chiclero* trails, newspaper articles chronicling *chiclero*-life, as well as academic studies on the history of the region's political history of forests. I am informed by all my previous research in the region since 2005 and will employ the regionally used term *chiclero* (tapper) and *chiclería* (referring to business and practice) to emphasize how this social history of the Maya Forest is commonly known and referred to. Consistent with the perspective employed in this chapter, neither maps nor exact locations of the biological stations are provided beyond the note that the three are within contemporary protected areas: One on the Mexican side of the Lacandon Forest Waterlands, one on the Guatemalan side of the Laguna del Tigre-Gulf Riverlands, and one on the Belizean side in the Caribbean Karst Waterlands delineated and described in the introduction.

A Hundred Years of Shared *Chiclería* in the Maya Forest Waterlands: Wayfinding with Go-Betweens

> Uaxactun, the name under which this site is known archaeologically, was not that first applied to it locally. When the aguado, or water hole, near the ruins (plate 53a) was first seen by chicleros in 1913, it was named San Leandro. In 1916, when the site was first visited by the writer (the Second C.A. Expedition), the paraje[1] was still known by this name but five years later in 1921, when the Fifth C.A. Expedition visited there, its name had been changed to Bambonal; indeed the name Uaxactun; by which it is now known at El Cayo, Benque Viejo and Flores, the three principal frontier centers of the chicle business in the Petén region, has come into general use only as a result of the Carnegie Institution's operations there.
>
> The names applied to the chicle camps of northern Petén are so fanciful, not to say trivial, such as El Triunfo (the Triumph), El Muerto (The Dead Man), El Cántaro (The Water Jar), etc, and they are so shifting, the same camp changing its name even from one season to the next that in all cases where new archaeological sites have been found near these ever shifting camps, the writer has felt justified in choosing for them new names derived from the Maya.
>
> Morley, 1938, p. 138

As the citation above shows, wayfinding and mapmaking sometimes work in mysterious ways. The archaeologist Sylvanus Morley did not take a liking to the *chiclero* names of places, which were the camps established by the chewing gum hunters, often located near important waterholes deep in the forest where the sapodilla trees could be found and tapped during the rainy season. The camps' locations were usually first sought by the hunters or foremen (*capataz*), who would explore the place with two criteria in mind: First, to have access to water (ponds, rills), and second, to have a number of good *chicle* trees within walking distance from the camp (Schwartz, 1990, p. 142).

Once the camp location had been found, the group of *chicleros* would move in and work in the area from July/August to January. *Chiclería* was a work of the rainy season, which was profitable for tapping the trees; the heavier the rain, the swifter the flow of the white gold. If the rain was too little, the latex would not flow, whereas too much rain resulted in an overflow, with wasted latex on the ground (Schwartz, 1990, pp. 143–144). The rainy-season work thus suggested cohabiting, quite literally, in *forest waterlands*: Slippery trees, flooded trails, and an excess of insects. However, prior to this watery experience, it was particularly the work of the *chicle* hunter or foreman to blaze the trails to-and-from the camp places, guided by the waterholes and a number of *chicle* trees, which also requires good background knowledge of the forest's ecology. As it turns out, the ancient Mayas may have been guided by similar principles, as ruins were often found by the *chicleros* near their camps, and to such a degree that sometimes the camping places were identified by the *chicleros* as *ruinerías* (places filled with ruins). Such was, for example, the region of Xmakabatum, described by the *chiclero* Julián Polanco, who discovered the site, as "a place of many ruins" (Morley, 1938, p. 423).

Image 5.1 Tapped Sapodilla Tree in the Maya Forest
Source: Hanna Laako, 2023

Of course, this may also be a matter of coincidence. However, since access to water is key for any civilization, it is no wonder that the *chicle* hunters and Mayas found the same water sources and thus connected with each other across time and space. At the least, Morley (1938) marveled at the Mayas' capacity to find their water access – in other words, what he found with the *chicleros*.

Nevertheless, Sylvanus Morley, an archaeologist from Pennsylvania, US, believed that the shifting *chiclero* names of places, which had essentially helped him arrive at the ruins, did not serve as adequate location names, and instead imposed ancient Mayan names that he chose, as the citation above shows. One might wonder whether this is colonial mapmaking of a political forest as discussed by Peluso and Vandergeest (2011), or Indigenous place-naming as discussed by Tuck and McKenzie (2015). It is difficult to say. However, clearly the *chiclero* places-in-knots and go-betweens mediating between the forest, their work, and visiting archaeologists in search of the lost Mayan ruins did not usually serve as worthwhile to use on a map as location names. This may, in fact, be interpreted as yet another wave of colonizing attempts. According to

Torras (2019, see also Kauffer, 2021), *chicle* history involved various stages of colonization. Torras (2019) explored *chiclería* in the case of Campeche, Mexico, where, in the first place, land titles were granted to foreigners, mostly North Americans. She refers to this process as *whitening* of the Campeche forest places, which later allowed North American companies to conduct *chicle* exploitation as the first stage of colonization. These processes were followed by settlers, who were required for the *chiclería* as laborers – migrants who arrived in different waves and became entangled in complex ways with the previous "natives". This colonization was based on exploitation of the workforce and involved mobilities in, to, and within the supposedly remote, that is, inaccessible Maya Forest places in Campeche, along the Guatemalan border. It was facilitated by states and governments that adopted contradicting and ever-changing policies, which often ended up placing the *chicleros* in historical invisibility. In other words, new colonizations were based on the total oblivion of the previous colonizations as cartographic illusions. This took place to such a degree that Torras (2019, p. 128) notes how in 2014, when a new Museum of Colonization, together with a statue representing the founding fathers of Candelaria, Campeche, was revealed, the first settlers were considered to have arrived at the place in the 1960s. This interpretation completely dismissed the earlier *chiclero* settlers, who had been pressing claims to have their history legitimately recognized for decades.

Chicle history also seems to be a passing line in most of the literature related to the Maya Forest region, although the oversight is not necessarily due to intentional amnesia. The oversight is due to many reasons, such as *chicle* often having been a secondary occupation for companies focused on timber trade, or it was preceded by processes perceived to have a heavier role in the regional economic history, such as *henequen, palo tinte* (i.e., Campeche logwood and mahogany), or it was followed by other businesses with contemporary importance, such as the *xate* palm. However, even these passing lines in the literature are important, because they help to unfold the historical *chicle* layer of the Maya Forest.

The studies that have focused on *chicle* history in the region have tended to emphasize its ties to foreign, capitalist exploitation, facilitated by the governments vacillating between disinterest, contradictory land policies, colonizing programs, and regulation of natural resources, which eventually left many *chiclero* settlements in what Lowenhaupt Tsing (2015) describes as capitalist ruins and amnesia (e.g., Konrad, 1987; Martínez-Reyes, 2019; Schwartz, 1990; Torras, 2019). Such authors as Torras (2019), Rodas (2014), and Mendoza (2023) also importantly point out that, as *chiclería* was a regional (transboundary) phenomenon with increasing competition for resources, it agitated governments to foster their international borders to protect their "own" timber and nontimber resources. The forest cycles and bordering walked hand in hand. Often, this meant further pressure to colonize the border areas, which later became entangled with land-rights issues, border conflicts, and pressures for conservation. The hundred years of *chiclería* included tempestuous times of emerging nation-states, with struggles over power and control within their porous borders.

Despite the history of capitalist and nationalist integration and exploitation, the *chiclería* does not only represent repression and victimhood. It is also a source of pride, roots, and knowledges that corresponds to the region's many communities, settlements, and histories (e.g., Pérez Aguilar 2014). In fact, for many *chicleros* and their families, the business initially promised freedom because it was linked to the decline of the old estates with their businesses, which left the laborers unemployed and from which laborers had recently escaped slavery-like conditions. In this mixed context of liberation and job hunting, the possibility of making good money while settling within the forest seemed like new freedom (Ponce, 1990).

The previously described *chiclero*-panorama coincided, first, with the quest for archaeological ruins, and later with conservation tendencies that represent the layers and complex entanglements of the *chicle*-originating social forest history, in which particular knowledge, trails, and roots play an important role.

While in his diaries (Rice & Ward, 2021), Morley extensively narrated his insect-bitten gloomy days and long rainy and muddy journeys – including the lost trails on several occasions – his descriptions on *chiclería* are narrow, submerged into the blurry background of the main storyline about a marvelous archaeological discovery. According to Schwartz (1990, p. 328), Morley even made a reference to the *chicleros* as "good-for-nothings", despite relying on them as forest guides. Moreover, the editors of his diaries, Rice and Ward (2021) noted the racism in Morley's writing. Yet, it must be said in his favor that, at least in his *Inscriptions of Petén* (1938), and despite giving principal credit of discovery for his Western fellow archaeologists, he clearly indicates that at least five out of his eight sites of discovered inscriptions were found by *chicleros*, who then later informed and guided the archaeologists to these places to be "discovered" by the latter. These five were the previously mentioned Uaxactun (located by an unnamed *chicleros*), Balakbal (located by the *chiclero* Vicente Esquivel), Xultun (located by the *chiclero* Aurelio Aguayo), Xmakabatum (located by the *chiclero* Julián Polanco), and La Honradez, which was

> discovered on January 11, 1910, by the Eight Peabody Museum Expedition under A.M. Tozzer, through information received from a *chicle* contractor, Eleutario Hernández, who met a violent death at the hands of the Colonies Police of British Honduras the following year for alleged revolutionary activities.
>
> (Morley, 1938, p. 432)

In this latter case of Eleutario Hernández, the ruins were actually named after his *chicle* camp, located two kilometers southeast of the ruins, by the Eight Peabody Museum Expedition and later confirmed by Tozzer: La Honradez. Mathews (2009, 89–90) also reported discoveries by *chicleros*, such as Francisco Morales, who discovered the site of Calakmul when cutting a trail through the jungle, and Carl Frey and John Bourne, who found Bonampak in Chiapas.

Of course, one can also wonder to what extent the archaeological sites were discovered by individual *chicleros*, who had acted as guides on the occasion that the archaeologist "discovered" the ruins. The knowledge on the location of the ruins may be collective in similar vein to other types of knowledges related to the Maya Forest area, as evidenced in Chapter 2 with the case of collective Indigenous medicinal knowledge and bioprospection.

Devine (2016) addressed the challenge in the case of the Ch'umil community located within the Maya Biosphere Reserve in northern Guatemala. According to Devine's (2016) interviews, this is previously mentioned by Morley as *Bambonal*, which he renamed as *Ch'umil*, which is the Mayan word for stars. The community of Ch'umil was established first as a *chiclero* camp by migrant workers, who had been working for the Wringley Company. When the *chicle* boom started to come to its end after the World Wars, many *chicleros* dedicated themselves to the collection of *xate* and as seasonal workers for the nearby archaeological sites, such as Mirador. The fall of *chicle* boom contributed also to the "looting fever" (*fiebre de saqueo/huechería*) in the 1970s and 1980s, when impoverished workers tapped into subsistence digging and looting. The looting started to mark the difference between licit and illicit archaeological findings in similar ways as today happens with wildlife hunting and trafficking, which often generates critique among communities: Looting became illicit for livelihoods, while archaeological permits allowed discoveries for researchers and heritage museums elsewhere. In a similar vein, contemporary

Image 5.2 Trail Used by *Chicleros*, Conservationists, and Archaeologists in the Maya Forest
Source: Hanna Laako, 2022

conservation has been criticized for categorizing the collection of endangered species for local people's livelihoods as illicit smuggling, while scientific collection of species for research is considered licit (e.g., Duffy, 2022).

As Devine (2016) notes, heritage and conservation work does not intend to perpetuate the unequal relations of power and wealth by categorizing certain activities as licit for the powerful that are illicit when done for local people's livelihood; however, this is often how it translates in the communities. Nowadays, critical archaeologists intend to address these inequalities, for example, by helping to establish communitarian museums – although criticisms still remain. For example, during my fieldwork, a forest guide with *chiclero* family history mentioned that while the archaeologists – mainly North Americans – regularly come to hire as many as 60 local villagers to work on the sites, thus providing important economic resources, when the artifacts are eventually found, "the local villagers are hushed aside and the artifacts just disappear. We do not know where".

Nearly a hundred years after its establishment, Ch'umil is home of almost 2000 Q'eqchi and ladino[2] residents (Devine, 2016). As the decline of the *chiclero* epoch drew many into economic struggles, resentment grew over the Mayan archaeology. According to the Ch'umil *chicleros*, the Maya ruins were never found by the archaeologists but by the *chicleros*, who often established the camps in the ruins given their access to water holes (Devine, 2016). Thus, the socioenvironmental history and heritage of the village is a combination of *chiclero* and Maya archaeology's landscapes. The community had established its own *chiclero* museum comprising Maya artifacts found by the *chicleros*, which was connected to their claim of treasures as their forest heritage based on their shared history of migration, *chicle* production, and archaeological excavation.

According to Schwartz (1990), who wrote an extensive *chiclero* history of Petén, the *chiclería* epoch lasted nearly 100 years, running approximately from the 1890s to the 1970s. In his view, it connected Petén, Guatemala, to global economic markets and affected the social history of the region so much so that he describes it as the epoch of formation of a forest society. This is because the business involved – directly or indirectly – whole families depending on *chiclero* salaries and/or otherwise impacted by the industry through family ties and economic relations within Petén's then population. His conservative estimate was that during the first half of the twentieth century, probably 50 per cent or more of Peteneros were directly dependent on *chiclería*. However, he also arrived at the conclusion that while *chiclería* as an extractive, nontimber economy played such a role in the social history of Petén, it did not change the society's social stratum and fabric. In other words, while the rich may have got richer thanks to *chiclería,* which may have also temporarily boosted regional economy and income overall, it did not improve the humblest livelihoods but rather widened the economic differences.

Schwartz (1990) observes that although mostly grouped according to one's origins to avoid unrest, the *chicleros* were multi-ethnic, composed of a wide stratum, including Peteneros from both Indigenous and ladino backgrounds, *milperos* [3] and cowboys, and foreigners from Central and North America – in particular, the migrant *chicleros* came from Belize, Tabasco, and Yucatán. Similar migratory processes took place elsewhere in the Maya Forest, such as in Quintana Roo. According to Mathews (2009), in 1918, large concessions of *chicle* extraction were granted to the new territories of Quintana Roo, which developed rapidly as one of the largest exports in Mexico and Guatemala. The growing *chicle* business benefited from the decline of *henequen* and logging in Yucatán. As *chicle* could not be well cultivated in plantations but grew in sparsely populated areas, large amounts of migrant *chicleros* were needed. While Schwart (1990) described Mexicans and Belizeans moving to Petén for *chiclería,* Mathews (2009), Martínez-Reyes (2016), and Pérez Aguilar (2014) note that also Peteneros and Belizeans moved to Quintana Roo. Similar exchanges and crossing of porous borders have also been detected between Peteneros and Campeche, and along the Usumacinta River between Chiapas and Petén (e.g., Mendoza, 2023; Rodas, 2014; Torras, 2019).

By the 1940s, the nontimber extraction business had grown so much that it ranked as the most important industry in Yucatan and the third most important in Guatemala, employing probably over 40,000 people (Mathews, 2009). Eventually, however, concerns were arising about the sustainability of extraction as it was calculated that careless tapping had killed 25 percent of sapodilla trees in Mexico by the 1930s. It was often thought that the local people knew to care for the trees so that they could be used again for tapping. However, the lure of the business and large number of migrant workers unfamiliar with the environment caused excesses.

The *chiclero* history writing has been typically gendered, mainly dealing with a masculine forest space and masculine labor history wherein the role of women is often reduced to cooks in the camps and described as subject to a promiscuous male gaze. Schwartz (1990) at least pays attention to the other end of the chain of *chiclería*, which is the domestic part at homes located in villages or communities. Mathews (2009) points out that especially during the final decades of *chiclería* in Quintana Roo, the men often took their wives and children with them to the forest. In addition, Berganza (2017) alludes to the active role of Belizean women during the *chicle* boom, and Torras (2019) cites female testimonies in Campeche. However, the *chicleros* tended to be young, free men who experienced the work as a rite of passage. Schwartz (1990) positions *chiclero* life as an opposite to the more established, communal *milpero* life. The *chiclero* life also comprises violent stories of aggressions and fights, although Schwartz (1990) considers them mostly exaggerated by the *chicleros* themselves. Nevertheless, by no means was the chewing-gum hunting life easy; rather, it was filled with occupational hazards like injury, sickness, harsh weather, and dangerous wildlife.

The commercial chewing gum was invented by John Curtis of Maine in 1848, when he began to use it from the spruce tree resin. However, from the 1870s onward, *chicle* became more important. It had its boom during the World Wars when provided for soldiers. The business started to decline shortly afterward when, around 1944–1945, it started to be replaced by synthetic gum. In 1866, the US manufacturers began tapping *chicle* in Veracruz, Mexico, and shortly afterward, the Yucatán peninsula became the main terrain for the extractive business. While *chicle* was being produced in various Latin American countries, the main area was Yucatán in Mexico, Belize, and Petén in Guatemala – namely, the Maya Forest. Indeed, it is fair to say that the *chiclería* integrated the full Maya Forest, not only for the shared business and migration but also for the interconnected routes. For example, the *chicle* produced in Petén had two main routes: It was transported whether by river through Tenosique, Tabasco, or by mule through Belize. Until about the 1930s, most of the *chicle* produced in Petén was smuggled out through Mexico or Belize to Canada and then to the US (Schwartz, 1990).

In his social history of *chicle* in Petén, Schwartz (1990) focuses on explaining the connections of Peteneros to global markets. In reality, the routes he describes are those that integrate the region of the Maya Forest as a shared forest-waterlands history. In his view, prior to the beginning of *chicle* history, Petén was wildly isolated – a hinterland of hinterlands, given that the neighboring areas in Mexico and Belize were also sparsely populated. The colonial Guatemalan governments had little interest in the deserted trails of northern Petén. In addition, the area covering Yucatán, Belize, and Petén was affected by both the Caste War that brought refugees and rebels also into Petén, which painted the northeast region of the Petén department as dangerous. Moreover, the western parts of Petén toward the Lacandon Forest Waterlands were little populated and rumored to be attacked by "wild Indians". These borderlands were constantly subject to gossip surrounding violence by *lacandones, San Pedro Mayas, nomads*, and *huites*.

Even during the *chicle* epoch, the region suffered from unrest and fear of violence due to the Mexican revolution, which also witnessed the entrance of different kinds of revolutionary groups and unsettled people. Sometimes, these were also *chicleros* protesting *chiclero* injustices. For example, in 1916, Mexican revolutionaries/*chicleros* protested in eastern Petén (Schwartz, 1990). The archaeologist Sylvanus Morley suffered himself from these hostilities, when in 1916, his crew was attacked on a trail from Petén to Belize, killing two crew members accompanying his expedition, while the other members, including Morley himself, barely escaped to the territory of British Honduras (Rice & Ward, 2021, pp. 300–314). The ambush was orchestrated by the Guatemalan troops, which were in the area apparently employed against Petén's *chiclero* and muleteer rebels. However, the events furthered the battles between the forces of British Honduras and the Guatemalan army. Indeed, according to Rice and Ward (2021, p. 249), the episode also caused diplomatic disputes between the US and Guatemala and between Guatemala and Great Britain. Evidently, Morley's future work in Petén was jeopardized; nevertheless, he was sent back to the field in 1917 as both an archaeologist and a spy for the Office of Naval Intelligence.

In other places, such as in Quintana Roo, *chicle* seems to have decreased hostilities between rebel Mayas and Mexican forces (Schwartz, 1990). According to Mathews (2009) and Martínez-Reyes (2016, see also Forero & Redclift,

2006), the Mexican government involved a Mayan leader in the *chicle* business, which eventually divided the Mayas. Martínez-Reyes (2016) suggests that *chiclería* worked to subjugate the Mayas to the control of the nation-state and capitalist system in the aftermath of the Caste War. In his view, *chiclería* was also one important node in the long chain of complicated relations between the Mayan people and foreigners in similar vein as these are also present in the layers of Archaeology and conservation.

During the *chicle* epoch, Petén–Belize trade relations increased as north-central Petén was considered dangerous and western Petén remained unexplored; thus, traditionally Petén was looking toward Belize. With the introduction of rubber tree (*hule*) tapping in the 1890s, the traffic through the San Pedro River to Tenosique also became important. The *chicle* routes were, of course, also connected to logging, which has a long history in the region. For example, mahogany trade began in British Honduras from the 1630s onward, having a boom between 1770 and 1840. Logging routes along the River Usumacinta were discovered around 1874. Both logging and *hule* later facilitated the *chicle* businesses and networks. However, the geographies were slightly different: Logging was focused on the Caribbean-oriented rivers, whereas most of the *chicle* trees were found in the north of Petén. While logging was more a dry-season activity, *chicle* was conducted during the rainy season.

Although the boom of the nontimber extraction came to its end, *chicle* tapping is still practiced in the Maya Forest. However, more than *chicle* extraction

Image 5.3 Rainy Season in the Maya Forest
Source: Hanna Laako, 2022

itself, *chicle* settlements, trails, and ecological knowledge are evident in today's biological stations and archaeological sites and are thus key to the formation of the Maya Forest (see also Pérez Aguilar, 2014). Indeed, the *chiclería* of the Maya Forest is comparable to the case of rubber-tapping in Amazonas bringing in new laborers to the rainforest. After the decline of the boom, these laborers rebelled for their rights as *seringuieros* against the expansive cattle ranching that came to replace rubber tapping in a similar vein to what happened in the Maya Forest (e.g., Torras, 2019). With the emerging global rainforest movement, a new alliance was built between conservationists and rubber tappers as *forest people* (see Chapter 2) who defended the Amazon rainforest from cattle ranchers. The famous *seringuiero* leader, Chico Mendes, died in the attempt (Smouts, 2003).

During my research on the Maya Forest between 2018 and 2024, I found that *chicle* history and trails quite literally tap into the geographies of conservation. For example, when interviewing coastal and marine conservationists and scholars in Quintana Roo in June 2023, the transboundary forest connections of the Maya Forest emerged in the discussion, pointing at the *chicle* history of Yucatan. Many of the communities that form the stakeholders of conservationists have settled as a result of *chicle* extraction in the 1950s and 1960s. These are run by Quintana Roo *chicle* cooperatives established in the newly pioneered *ejidos*, that is, communally held lands based on tenure (e.g., Martínez-Reyes, 2016; Sosa et al., 2022). The *chicle* references also emerged in the autonomous communities in *Cañadas* microregion of the Lacandon rainforest in Chiapas, established as a result of *chiclero* explorations (Laako's interview with a community member in March 2024). The *Chiquibul* National Park in Belize and *Chiquibul*-Maya Mountains derive from *chiquibul*, a particular *chicle* tree found in this region. In the surroundings of the Belizean-Guatemalan border, El Cayo, today called San Ignacio and its surroundings, is filled with *chiclero* histories. In the biological-archaeological *Milpa Lodge* station of Programme for Belize, in the Rio Bravo Conservation and Management Area, a *chicle* trail emerges with old, scarred sapodilla trees. The Milpa Lodge is located right on an ancient site of a *chiclero* camp. These are only a few examples.

However, the objective of this chapter is not only to point out the formative role and socioenvironmental importance of *chicle* to the Maya Forest but to address the ways in which the conservationists, now located in the *chicle* places or nearby them, relate to the contemporary dynamics and challenges of the Maya Forest as the new go-betweens and curators of the Maya Forest waterlands.

Biological Station 1: Curating in Chiquibul Biocultural Borderlands

It was an early and somewhat gloomy, rainy morning in Flores, Guatemala, when we – my colleague and I – left our hotel and headed for Melchor de Mencos at the Guatemalan-Belizean border in June 2022. Despite the Friday-morning rush hour, and still facing COVID-19 restrictions, we managed to cross to Belize quickly and took a taxi to a quiet, tourist border town called

Succotz for an interview with a conservation organization. However, when arriving in Succotz, the town built along the Mopan River hillside, we realized we were too early and thus headed for breakfast, where we were reached by a patrol of the Belizean armed forces who kindly asked about our business in Succotz.

After our interview with the conservation organization, our journey continued toward the Chiquibul National Park research station. Passing El Cayo (today called San Ignacio), we headed for the Pine Ridge Forest Reserve and then crossed Chiquibul Forest Reserve toward the Caracol Archaeological Reserve, before turning toward the research station where we passed a couple of military and ranger checkpoints. Our driver was somewhat nervous as he had not driven to the station before and was not sure about the route. In addition, the heavy rain had made the road muddy, with the heightened risk of getting stuck, even on four wheels. What we witnessed was a transition from pinewoods to rainforest under heavy construction: important road building was taking place to ease tourists' access to the Caracol Archaeological Reserve.

After some bumpy hours, we reached our destination, where we spent the weekend. Composed of several wooden houses on stilts in a typically Belizean style and surrounded by the hot, noisy, insect-filled rainforest, the station was all that could be imagined for research and patrolling: It was connected by radios and other equipment and included a library with maps and other sources, as well as rooms for lecturing. Indeed, we shared the place with a lively, energetic group of North American biology students and their professors, with whom we conversed over the weekend when coinciding with them in their walkabouts to check camera traps and species. Again, we were guided to observe the trails, which happened to be those of earlier *chicleros*, as our guide pointed out on several occasions.

As suggested by Morley (1938) and Schwartz (1990), the Petén–Belize borderlands became important places-in-knots for *chicle* routes and trails, later occupied by archaeologists and others. First, the node composed of Benque Viejo and El Cayo (today, San Ignacio), both of which were born as lumber camps, and which often also includes San José Succotz (a small town established by Yucatec Mayas escaping the Caste War), connects the Caribbean shores of transportation by the River Belize with its tributaries and provides access to northern Petén-Belize *chicle* locations, which also include many of the archaeological sites narrated by Morley (1938). He mostly conducted his expeditions first by coastal ride from Quintana Roo via Belizean coast and then from the node Benque Viejo/Cayo/Succotz toward Petén. Until the 1940s, the El Cayo node could be reached from Belize City only by the river. Hence, the node was first established by logging companies that set their camps along the River Belize, which was then used by the *chicleros* at the end of the nineteenth century, when the mahogany business began to wane. Nowadays, this node is connected by Melchor de Mencos – that is, the access point to the Guatemalan side, which has also included special operations forces of the Guatemalan army.

Image 5.4 A Wall Painting in San Ignacio, Belize, Illustrating the History of Logging and Rivers
Source: Hanna Laako, 2023

According to my interviewee from the Belizean Institute of Culture and History (Belmopan, June 2023), the *chicle* epoch represents a golden era of Belize, particularly for the socioenvironmental history of this place-in-knot comprising today's Benque Viejo/Cayo/Succotz: The usual saying was "the common laborer would wrap his cigarette in paper money". Women would work as cooks in the camps or in the milpas at home; however, the *chiclero* brought the main income. The trade was initiated by the Lebanese and Turks in Belize, and as trade unions were not recognized, it led to exploitation, slavery, and resistance (see also Roessingh & Darwish, 2012). However, the trade augmented connectivity across the porous border, creating a node of its own, and brought about a vast amount of ecological knowledge related to trails and water holes, weather, plants and animals for survival, and medicinal herbs. Berganza (2017) describes how the industry allowed the El Cayo node to flourish and created connectivity given the possibility of self-employment, which was considered to mark an important difference compared to the earlier "colonial misery". According to my previously mentioned interviewee, the *chiclería* also expanded territorial knowhow as it was not dependent on the rivers, like logging. On the contrary, the *chicleros* needed to abandon the riversides and enter the vast forests in the rainy season to find the water holes for camps and suitable locations for the sapodilla trees (see also Awe, 2004 and Tzul, 2004).

Given that the *chiclería* epoch lasted nearly a hundred years (according to my interviewee, it started to erode in the 1980s in Belize), it was also tied to the other dynamics of the political forest by being defined by the Belizean-Guatemalan borderlands. While the Wyke-Aycinena Treaty of 1859 established the borderline – officially called the adjacency line – that is currently used to mark the international division, which is contested by Guatemala and submitted by both states to the International Court of Justice, the borderlands have been mobilized in many ways. Among others, the area has been subject to arms smuggling through the rivers to rebels of the Guatemalan and Mexican side during the Caste War, and later involved competing houses that engaged in patrolling at the beginning of the twentieth century. The Guatemalan civil war witnessed disappearances of people in the Belizean node of Benque Viejo/Cayo/Succotz, but the area also received many refugees.

According to Matola and Platt (1998), nearly 80% of the Belize–Guatemala border forms an active corridor for wildlife species migration; thus, the political forest is also an ecosystem that owes its name to both the sapodilla tree, called *chiquibul*, and the *Chiquibul River*, which flows from Belize into Guatemala and back to Belize, forming the headwaters of the Belize River. The Chiquibul National Park lies in the largest karst area of Belize, which incorporates parts of the Chiquibul Cave System. The Chiquibul Forest Reserve was originally established in 1956. However, in 1991, due to lobbying by conservationists, the reserve was partially reclassified by separating two new areas: The Chiquibul National Park and the Caracol Archaeological Reserve. According to interviews carried out in Belize in 2022, the Chiquibul forms one of the three main conservation priority areas in Belize, in addition to the coastal reefs and northwestern forests.

The history of the Chiquibul area and the research station currently located there reflect the turbulent times of the borderlands described above. According to my interviews when I first visited the place in 2022, the Chiquibul borderlands are characterized by governmental disputes that "colonialism left behind". Prior to the 1990s, the reserve had forest concessions, but with the absence of forestry law enforcement, extraction became wild. Between the 1980s and 1990s in particular, the transboundary area was characterized by many illegal activities, among which was poaching, the extraction of precious woods and gold by inhabitants entering from the Guatemalan side communities. On the Belizean side, no villages were built within the Chiquibul; rather, the perception is that the Guatemalan communities often consider the Chiquibul lands as theirs, given the intergovernmental disputes. However, the Guatemalan villagers also enter the Belizean side of the Chiquibul to alleviate the economic hardship and due to the lack of natural resources in the deforested communities on the Guatemalan side. The entrance of Guatemalan villagers has, again, caused several crossfire incidents with the Belizean armed forces, leading to a loss of lives, and the diplomatic relations between the countries have tensed further.

The research station in the Chiquibul forest was first founded as a *chiclero* camp, which was used during the rainy season running from July to February by *chicleros* that entered the area from the El Cayo node. In the 1970s, the camp seems to have been used for hunting and occupied by a family that eventually abandoned the place. In the 1980s, the camp was recreated through the efforts of the Belize Forest Department and the Natural History Museum of London. In the 1990s, it was developed toward the idea of a research station. However, these were unsettled times: Looting, smuggling, illegal logging, and expansive cattle ranching across the border occurred, which made it difficult to maintain the place.

The station was abandoned due to unrest several times until 2007 when, after several attempts and speculation concerning the Chiquibul situation, it was acknowledged as a security problem for Belize. In 2007, a nongovernmental organization from Succotz signed its first co-management contract in Chiquibul National Park. The organization was established at the end of the 1980s as a youth group interested in the outdoors. In 2015, the organization took on the station's administrative duties, developing it as a place for research, which nowadays receives students and scholars from many North American universities yearly. It is surrounded by, and is in active collaboration with, many military camps, including the one in which members of the British royal family have had their military training.

According to my interviews in Succotz in 2023, the conservation organization has managed to curate the Belizean Chiquibul forests: In their view, now the station can receive both international and Belizean visitors, including student groups, to learn and enjoy the place. Near the station are sites that have been used for teaching students in forest survival skills and species. Back in the 1980s and 1990s, the Chiquibul was perceived as a dangerous spot on the Belizean

Image 5.5 Biological Research Station in Chiquibul Borderlands, Belize
Source: Hanna Laako, 2022

map, a place where “nobody would have sent their kids” (Interview with the conservation organization, Succotz, June 2023).

However, this curating has also required more far-reaching transboundary and diplomatic efforts. According to my interviews conducted on both sides of the border between 2022 and 2023, conservationists agreed that the transboundary Chiquibul needs to be considered as a shared borderlands, not only for its shared historical developments but also for its contemporary challenges, particularly related to the economic struggles in the Guatemalan side communities that enter the Belizean Chiquibul National Park in search of endangered species (such as the red macaw) for smuggling, precious woods, and space for cattle ranching. It seems that, lately, drug cartels have attempted to do the same. The Chiquibul is a transboundary, socioenvironmental, and interdependent landscape and a shared ecosystem that requires a bilateral element and binational counterparts in a biocultural landscape focus. A Guatemalan conservationist characterized the transboundary Chiquibul borderlands as “integrally Indigenous lands”, whereby the Guatemalan state had “socially neglected the population” and the Belizean state had “neglected the rights of its Indigenous”, while “the (ancient) Mayas did not recognize or live by this border” (interview in Petén, June 2022). In other words, for the interviewee, the border divides the Maya people’s traditional homelands, while neglecting them on both sides in different ways.

According to the Guatemalan conservation organization, and the Guatemalan Council of Protected Areas CONAP (2019), the communities in the Montañas Maya-Chiquibul Biosphere Reserve include, in particular, Mopan Maya and Q’eqchi people in addition to ladinos, many of whom also took active part in the *chiclería* and continue to practice the extraction of nontimber products, such as *chicle* and *xate*, some of which are collected from the Belizean side. In addition to sacred cave and river systems, the area also includes over 50 Mayan ruins, such as Sacul and Ixtonton as the main ones.

Given the border tensions between Guatemala and Belize, the governments on both sides – according to the conservationists – have been careful not to increase hostilities and tensions. For these reasons, the Ministry of Foreign Affairs of Belize, and its counterpart in Guatemala, have granted the two conservation organizations and their partnership allowances and law enforcement entitlements to alleviate the socioenvironmental challenges within the transboundary Chiquibul area.

Thus, since 2007, the conservation organization co-managing[4] the Chiquibul area in Belize, which has been extended to other protected areas along the borderline, has made a transboundary effort to improve the livelihoods in the villages on the Guatemalan side to reduce border-crossing incidents. This has been done by establishing an alliance with the other conservation organization from Guatemala. Additionally, the Belizean organization is collaborating with the Belizean armed forces in patrolling and controlling trespassing activities.

On the Guatemalan side, the conservation organization points out that the socioeconomic problems in the villages along the Belizean border are well acknowledged. In their view, the Guatemalan Chiquibul Montañas Maya

Biosphere Reserve was created in 1995 without consulting the villages located within it. These were the product of an earlier policy of colonization by the Guatemalan agricultural institution FYDEP, which was later replaced by the Council of Protected Areas (CONAP). Thus, the Guatemalan conservation organization notes a reverse panorama in which the Belizean conservation organization must collaborate with the Guatemalan side to conserve Belizean landscapes as the rates for illegal logging and smuggling have been "high".

As a result, the Belizean conservation organization has promoted a Chiquibul Peace Park[5] between Belize and Guatemala; however, given the border disputes, this has not materialized. Instead, the two organizations have been given governmental allowances to attend to the socioenvironmental problem corresponding to the transboundary Chiquibul. The steps taken so far have been so successful that the work has been awarded with peace prizes. At the level of the contested forest waterlands, the situation is not always so easy. Both organizations are aware that, in addition to nature conservation, managing socioeconomic conditions and keeping the tense balance in the border have been outsourced to them, nongovernmental organizations. On the Belizean side, the organization notes that they have been criticized about for curating Guatemalans when more important problems are yet to be solved in Belize. On the Guatemalan side, again, the Belizean organization has been occasionally tagged as a paramilitary organization by those who are not happy with its patrolling and law enforcement.

The Chiquibul curating certainly extends from fragile ecosystems to livelihoods and geopolitics – in this case, with the attempt to consider places-in-knots beyond borders, characterized by people traveling pathways that have survived the complex transboundary histories. According to Saxer (2022), curating is different from extractive mapping, which more resembles collaborative gardening. In his view, conservation does not work without local sensitivity, which relates to cultural identity, history, and belonging. However, as is the case of the Chiquibul borderlands, the layers of biodiversity and cultural heritage are often left out of issues of security and geopolitics in the context of a contested border and a postcolonial small state such as Belize (Kauffer, 2023).

Biological Station 2: Go-Betweens in the Contested Biocultural *Laguna del Tigre* Lands

Just before the Easter break in April 2019, we arrived at a small town located in the borderlands of Balancán, Tabasco, and Petén, Guatemala. We had been in the area conducting interviews on conservation, among other topics, and wanted to get a glimpse of the border region between Balancán and Petén. This had been easier said than done, as we were told that the area was in the hands of a Mexican drug cartel and that we definitely should not move about there after daylight. During the earlier days, we had been driving there, conducting interviews, and observing bush fires and cattle fields suffering from lack of water at the end of the dry season.

Then, our guide, a beekeeper, happened to remember that she had a beekeeping friend whose family lived in a town near the Guatemalan border and could receive us. Beekeeping has become an important conservation measure in the deforested Tenosique-Balancán borderlands (e.g., Vera-Martínez & Ceballos-Falcón, 2024). We were kindly invited by the beekeeping family for a delicious mole-lunch in their home, which is also when we realized that the retired father had been a jaguar hunter and a smuggler of precious woods in the 1980s. Although our attempts at generating trust for such a chat turned out to be unfruitful due to their suspicions, the family agreed to accompany us to visit the border instead. Hence, we were packed in our fieldwork-Hilux with the extensive family and proceeded to drive the short, bumpy distance to the Balancán-Petén borderlands.

What we observed when walking along the border was a vast, bushy landscape extending endlessly toward the Laguna del Tigre National Park of Guatemala, accompanied with some border posts filled with shotgun holes. The father acknowledged that the jaguar frontier was far away on the Guatemalan side – no more jaguars or other big game in these bushes. The lands we were stepping into were part of the nuclear conservation area of the Maya Biosphere Reserve and a Ramsar site, designated to protect its complex seasonally flooded forest, slow-flowing rivers, marshes, permanent lagoons, and seasonal water bodies – that is, Guatemala's largest internationally recognized wetland (Ramsar, 2024).

The following day, we had yet another excursion, this time by the transboundary San Pedro River, which crosses both previously mentioned border areas. At the time of our research, attempts were being made by the Tabasco side to designate the river also as a Ramsar site on the Mexican side. In our interviews, we had perceived the excitement over the news about the advancing Maya Train project, which was going to connect Yucatán's tourist routes to Balancán and Tenosique in Tabasco. This generated hopes for converting these deforested Tabasco riverlands into the tourist attraction it had once been in the 1980s, when "the *mochileros* (backpackers) came from Palenque and survived only by eating bananas". We were shown Mayan ruins still to be uncovered and a heron sanctuary island in the San Pedro River, which, unfortunately, had been cleaned from the herons by smoke because the nearby community wished to polish the beach for Easter-break tourists.

The plan was, however, to travel upstream the San Pedro River, not only to reach as close to the border as possible, but also to access the lagoons that contain a specific type of Yucatán mangrove that the researchers have explored to understand its existence so far away from the peninsula (e.g., Fiscella, 2020). We were accompanied by two guides: One, who was sober and serious and stuck to his navigation task, and another one, who had prepared for the trip with a bottle of *aguardiente* [6] and, quite shortly afterward, was more in charge of the entertainment than navigating. Yet, despite the good spirits, we could not advance much closer to the border or see the mangrove, due to the dry-season effect that blocked our possibilities to advance further by boat – the water was not enough.

Image 5.6 San Pedro River Pictured from the Biological Research Station Located in the Laguna del Tigre National Park, Guatemala
Source: Hanna Laako, 2022

Fast forward to June 2022, when we were again on a boat along the San Pedro River, only this time on the other side of the border, in the Laguna del Tigre National Park, Petén. Prior to this particular moment, we had met up with a conservation organization in Flores and accompanied them by car down a bumpy road, crossing villages of the Maya Biosphere Reserve to the final town, where the San Pedro River begins. In this spot, we had boarded a boat to take us to a biological station, which had been run by the conservation organization since 2008. The station was built in 1994 on a site by the San Pedro River called the *San Rafael chicle camp*.

Indeed, in the forthcoming days, when we hiked and rode along the San Pedro River, what we observed a forest filled with *chiclero* trails occupied by conservationists, archaeologists, and community members circling on those muddy routes – some of them walking, like us, and others driving four-wheel jeeps. The forest's busy hour became particularly visible when we hiked toward the archaeological site of Perú. We only managed to arrive at what is called the Perú Camp, which is still four kilometers away from the scientific station called Laguna El Perú and in close proximity to the Perú-Waká ruins. However, even the Perú Camp was worth experiencing with the archaeologists camping there and community women busy cooking *tortillas*. The place was filled with signposts, as if prepared for any tourist that might stop by in what, in our eyes, seemed a remote and out-of-place area, but that clearly was an important northward place-in-a-knot. Indeed, the San Pedro River has been an important route of access to the northern Petén Forest. According to Dugelby (1998, p. 161), the Maya Biosphere Reserve comprises roads built for logging and oil exploration. However, these are accompanied by thousands of footpaths and trails that serve to transport men and supplies in forest camps, which are scattered throughout the reserve, each located adjacent to a water supply. Many of these trails used for *chicle* harvesting are not passable by cars during the rainy season, hence the reliance on mules.

The biological station itself has a turbulent history. Between the 1970s and 1990s, when the station was still known as San Rafael, it worked as a "site for extraction" (interview in the Biological Station, June 2022). While *chicle* was the principal source of extraction, the camp was also used for the logging of precious woods, looting of archaeological pieces, and smuggling of wildlife, such as the red macaws. Prior to 1975, the camp could be accessed mainly by an airstrip, which was later used in the civil war by the guerrilleros and military. However, the use of this airstrip also left the nearby town "free" for decades, until its contemporary growth due to migration and demographic explosion. According to the local member of the organization, given the camp's history as an important place for extraction, it has always been controlled by or has been of interest to "coercive actors, peoples with money". Indeed, in the 1990s, it

Image 5.7 Trails and Campsites in Laguna del Tigre, Guatemala
Source: Hanna Laako, 2022

became a center of activities related to extraction of precious woods, archaeological looting, hunting, and wildlife trade.

As a result, another conservation organization, created in 1991, decided to build a station in San Rafael that would be strictly scientific, with the aim of fostering conservation toward the southeastern parts of the newly established park. In the process, the camp ardently changed its name, although this was by no means an easy task given that the place was firmly known as San Rafael. As a result, according to my interviews, the communities started to organize in terms of their extractive activities, arguing that they would not survive by conserving. As a result, in 1997, groups entered the biological station, burned the installations, and took the personnel as hostages.

My interviewee was in a nearby landing when the events unfolded. He told me that boats with people had passed by him first, asking about the whereabouts of the biological station. Since he had not been aware of what was going on, he had indicated with his hand that the station was only a short distance upstream. With a dry laugh, he said that he had not realized what was happening until he had sensed the smoke coming from the burned installations, and a while after, the same boats had passed him again – only, this time, they had also been carrying his mates from the station, sitting on them with their hands tied. In his view, it was later clarified that with the hostilities, the attacking groups demanded the legalization of the lands within the park. While he believed that the biological station was not responsible for land distribution, he acknowledged that the protected areas were created without considering the existing villages in the area. In other words, "a lot of people were affected by the creation of the protected areas, and hence, the hatred toward the Guatemalan Council of Protected Areas (CONAP) augmented" (interview in June 2022).

In his view, however, the criticism was not only addressed to the Guatemalan governmental institutions, but it was rather thought that, through them, "the *gringos* were stealing from us" – perhaps not only related to civil war and contemporary conservation but to the region's prior economic background, as evidenced in the trails and camps history. Yet, as a result, conservationists have become positioned as complex mediators and actors between the contemporary extractive and war history of Petén and the peasants' land-rights struggles, which also connect to the contemporary extractive stories in places-in-knots.

The 1997 events in the station have contributed at least to two different but entangled narratives. One takes the peasants' viewpoint and is focused on green land grabbing, in which the contemporary drug-cartel problem is also perceived as a tool to criminalize peasants (e.g., Ybarra, 2018). In this scenario, conservation is perceived as an ally for the militarization, foreign involvement, and land-right injustices of contemporary Petén. Another scenario is that of the conservation organizations that point out the problem of criminal groups and their coercion of peasants.

Nevertheless, my interviewee believed that conservationists had also changed their approaches since the conflicts unfolded at the end of the 1990s. Now, conservation work is perceived as something that needs to contribute not just environmentally but also socially. In his view, the conflict related to conservation derives from the contemporary history of migration of communities escaping from urban contexts to rural ones to be then trapped by the creation of protected areas. As these tendencies make conservation impossible, organizations have aimed at integrating local members and/or leaving the organizations in the hands of the local people. In this way, he concluded, conservationists are making an arduous attempt to *resignify* the park, the forest.

The background of this conservation-organization outsourcing to local hands is, of course, related to the severe criticism of foreign involvement in Petén, which accentuated in the 1990s with the creation of the Maya Biosphere Reserve and was viewed as continuity for other type of previous foreign involvement (e.g., Rodas, 2014; Ybarra 2018). In the 1990s, the different protected areas within the biosphere were mostly financed and managed by international conservation organizations (e.g., Millner et al., 2019; Taylor, 2010). As conservation was politicized, many organizations withdrew or aimed at involving local actors. According to my interviewees, who have always been local in their majority, the international organizations also justify this responsabilization by arguing that for the future of conservation, the presence of international organizations cannot be trusted as, in some situations, they may need to retire, and therefore, only the local people stay. This entails, of course, another discussion related to the complexities of local people who cannot leave conflict zones.

Nevertheless, such authors as Rahder (2020) also provide a slightly different viewpoint. According to her research, conservationists, in fact, now tend to avoid such strictly conserved areas as *Laguna del Tigre* for its violent, unmanageable landscape and rather prefer to *collaborate* with the community forestry concessionary area, which has also received foreign funding.

In 2008, the biological station in question was taken by another local nongovernmental organization established in Petén in 2001. By then, the station had basically "come down" – it had no maintenance. Between 2008 and 2014, the station was completely remodeled and focused on ecotourism that would engage the nearby villages. In the station, a considerable effort was made in creating hides and possibilities for wildlife observation and trekking the "Maya trails".

The conservation organization is mediating between various contemporary Maya Forest dynamics. First, the organization is engaged in the strategic bilateral alliance in the Chiquibul borderlands described in the earlier section. Second, it is engaged in conservation efforts in the Laguna del Tigre borderlands and southward toward the Sierra del Lacandon borderlands, which are currently affected by drug-cartel and criminal-group activities. According to Devine et al. (2018), the Laguna del Tigre National Park is an example of a political forest with changing land use by drug-trafficking organizations through *narco-ganadería* – narco-cattle ranching that involves money-laundering practices and deforestation resulting from the attempt to control territory and drug routes using forest places to build discreet landing strips. In their view, the drug trafficking expanded in the Tabasco–Petén borderlands in the early 2000s, when Mexico adopted a new, militarized approach to the smuggling routes, which then contributed to transitioning the smuggling to Peten's large protected areas perceived as strategically fit: The waterlands served for cattle ranching, which again contributed to money laundering. Additionally, the lands were sparsely populated, and the existing villagers could not denounce the activities. According to Devine et al. (2018), these factors contributed to the impunity of the narco-cattle ranchers, in addition to the existing policies favoring expansive cattle ranching, and the existing smuggling trails across the border. Thus, the Tabasco–Petén borderlands have witnessed a narco-landgrabbing that engages with the intimidation and violence of local people, who are also criminalized for the pending land issues.

As most of the Laguna del Tigre lands are deforested, a third sphere of discussion has emerged: Who is better at conservation? The conservationists in the deforested Laguna del Tigre or the neighboring communities of the multiple-use zone of communitarian forestry concessions that have managed to maintain more forest coverage? The Association of Forest Communities of Petén (ACOFOP) that represents the forestry communities of the multiple-use zone points out to the same socioeconomic history of Petén: The *chicle* extraction and foreign logging concessions that reduced biodiversity in the region until the transition of the 1990s, with the entrance of new conservation and protected areas as the "ideal". In the viewpoint of my ACOFOP interviewee (Flores, Petén, June 2022), conservation was perceived by the peasants as a landgrab by foreigners that clashed with the ongoing land distribution process at the end of the Guatemalan internal conflict (1960–1996). In other words, the reserve was created despite the settlements that had existed there for nearly 100 years due to the epoch of the *chiclería*. Thus, the ACOFOP was conformed mainly by *chiclero* cooperatives and communities (some of which also had collaborated with *chiclero* communities in Quintana Roo in communal forestry pilot programs) to defend the forest and *chicle* settlements found within (see also Millner et al., 2019; Taylor, 2010). As a result, the forest has been reorganized as the multiple-use zone of communal forestry concessions that today perceive that their form of conserving the Maya Biosphere Reserve is better than that of the Laguna del Tigre. According to Sundberg (2003), the contemporary history of the multiple-use zone comprising community forestry has created a new, democratic actor: That of the forest concessionary, born of the contemporary migrant history related to *chicle* settlements.

Conservationists are, once more, embedded in the roots and trails of the *chicle* camps and settlements and their global knots that have shaped the forest and its dwellers and their frames of interpretations as to whose hands the place is in and whose it should be in, before falling into the hands of drug cartels. In the case of Petén, the way in which *chicle* history is referred to seems also to carry a certain pride and continuity, and according to Rahder (2020), conservationists are, to some extent, in dialogue with that history in the sphere of community forestry.

Biological Station 3: Cartographic Illusions in the Biocultural Lacandon Rainforest

In March 2024, I was engaged in a vivacious online interview with a long-time member and activist of the Organization of Indigenous Doctors of Chiapas (OMIECH) established in 1985 by Tzotzil, Tzeltal, and Ch'ol health promoters and doctors of traditional Mayan medicine. My objective was to ask about his experiences and insights concerning the bioprospecting case of the Maya International Cooperative Biodiversity Group (ICBG) and the Traditional Indigenous Doctors and Midwives Organizations of Chiapas (COMPITCH, related to OMIECH) that was politicized at the end of the 1990s in Chiapas, Mexico, and mentioned in Chapter 2.

Although we did discuss the particular case and its implications, I was suddenly struck by his comment about his community, located in the microregion *Las Cañadas* in the Lacandon rainforest, right at the border of the Montes Azules Biosphere Reserve, established in the 1960s by his grandfather, a *chiclero*. Indeed, my interviewee explained that his grandfather had worked as a *chiclero*, and his work had consisted of finding the adequate places located by the waterholes. His wife had worked as a cook for the *chicleros*. After they had abandoned the *chiclería*, they had sought the excellent place the grandfather had previously located to establish their new lives there. Later, they were followed by others, converting the settlement into an *ejido*.

This is a remarkable detail, as *chiclería* as such is not often narrated in the numerous scholarly studies focused on the Lacandon rainforest, despite the forest being filled with sapodilla trees (e.g., Carabias et al., 2015). *Chicle* is mentioned in the passage by numerous studies and yet seems to vanish under other issues. Currently, it is difficult to dig into the *chiclería* in the case of the Lacandon rainforest, particularly when linked to archaeology and conservation. This is due to the extreme politicization and fear that has made people and organizations weary of indicating anything that might or might not be found in the rainforest that could incite further interest in the perceived resources. Conservationists may fear that indicating such natural resources as *chicle* might result in further settlement pressure and extraction in protected areas. Communities may fear that indicating such findings as ruins or species might result in further conservation. Organizations and governments may fear that any of these might worsen the uncontrollability of a region that already has a history of insurrection. One might ask: Does the extreme politicization of forests result in the conscious effort by local actors in the purposeful reproduction of cartographic illusions – that is, a space that is emptied of life/resources due to fear?

Now, however, I have jumped ahead to my story about *chicle* history and its whereabouts in the Lacandon rainforest. This supposed silence does not mean that this history does not exist, but rather, it hides behind the shadows of logging history first and in the obscurity of the contemporary layers of Lacandonia's political geography and history second.

In the case of the Lacandon rainforest, it was mainly the *madereros* (foresters) and *fincas* (rural farms or estates) that engaged in *chiclería* as a secondary occupation. The great historian Jan De Vos, in his much-cited book *Oro verde: La conquista de la Selva Lacandona por los madereros tabasqueños 1822–1949* (first edition 1988, 2015), mentions *chicle* only a couple of times. First, to limit the scope of his study, he mentions on page 10 that he will not address secondary topics, such as the extraction of *chicle* in the northern parts of the rainforest in the 1940s. Later, he mentions (2015, p. 64) that, by the 1870s, the exploitation of precious woods in the border areas with Petén had become so important that taxes started fall due to the companies and that these could also be paid in *chicle*. At the end of the book (p. 298), he mentions that when the lumber companies were accused of peasant exploitation, the governors of Chiapas and Tabasco denied the existence of *monterías* or *chiclerías*, although some "cutting work" was said to have taken place in the borderlands of the two Mexican states. These data are confirmed by Legorreta (2008) and Mendoza (2023), who sustain that, in the 1940s, the Lacandon rainforest experienced a *chicle* boom involving the estates that built small airstrips in the forest to fly the latex out because adequate roads for transportation did not exist. These *chicleros* – mainly outsiders who also adventured in the Guatemalan side forests – occasionally encountered hunters and archaeologists. However, by the end of 1940s, the business dried, and the trails vanished. Indeed, according to De Vos (2015), in 1949, when the Mexican government prohibited the extraction of unprocessed mahogany, and the final company withdrew from the rainforest, the remaining loggers had two options: To become *chicleros* or laborers in the *fincas* or ranches in Ocosingo and La Libertad in Chiapas, or Tenosique in Tabasco. In his other famous book, *Una tierra para sembrar sueños: Historia reciente de la selva lacandona 1950–2000* (first edition 2002, 2015), De Vos mentions that the history of the "deserted Lacandon rainforest", occupied only by the *caribes* (the lacandon), *monteros* (loggers), and *chicleros*, ended around the 1950s and 1960s, when the land was occupied by "guerrilleros and peasants".

Indeed, the limited pages of this chapter are not sufficient to narrate the complex history of the Lacandon rainforest colonization and posterior events in detail. However, to understand my interviewee's *chicle*-based community history, it is worth examining, as he suggested, the work by Leyva Solano & Ascencio (2002). In *Lacandonia al Filo del Agua*, the authors refer to the Lacandon rainforest (or Lacandonia) as a shared, heterogeneous socionatural unit that is open and flexible but also subject to its own particularities, especially within its many microregions. The authors (2002) sustain that while the northern parts of the Lacandon rainforest were mostly colonized by peasants from the 1960s onward due to the retirement of logging companies, as mentioned by De Vos, many other parts of the rainforest had already been populated from the 1930s onward by peasants from the Chiapanecan *fincas*. These peasants escaped the harsh conditions of the estates, first occupying the nearby forests and then entering deeper into the "Lacandon desert". Based on oral histories, Leyva Solano & Ascencio (2002) essentially call this the "exodus", suggesting an extensive escape from a *finca* frontier that had been formed at the edge of the Lacandon rainforest, circling the areas of Rivers Santo Domingo and Grijalva in southern Chiapas and passing east of the city of San Cristóbal de las Casas toward Palenque. Some of these early colonizers were originally from other states of Mexico but opted to stay in the rainforest instead of returning to their places of origin after leaving the *fincas*.

In effect, some of these early inhabitants were *chicleros*, who were also known to have blazed the trail for others after them, given their role as guides and trackers that helped many Indigenous groups and peasants to find available land (Leyva Solano & Ascencio, 2002, p. 46). Those who followed did so for many reasons, such as the situation in their places of origin and the agrarian reform that encouraged peasants to find new land for *ejidos*. Thus, the new rainforest inhabitants were peasants not only from Chiapas but also from many other states, with most being from Tabasco, Veracruz, Oaxaca, Campeche, Guerrero, Puebla, Mexico City, Michoacán, Yucatán, Estado de México, and Quintana Roo. In the case of the Lacandon rainforest, these factors together resulted in the accelerated population growth within the forest, especially from the 1950s and 1960s onward, given that the occupied lands started to become saturated, and the colonizing movement actively motioned toward the inner parts of the forest. Thus, the forest fringes in the north, around Palenque, and in the western parts were inhabited either by *chicleros* or by peasants escaping the plantations, guided by *chicleros* or their trails.

Prior to these *chiclero*-related mobilities, the forest was mainly occupied by the Lacandons, originating from the Yucatán during the colonial era, who had lived mostly dispersed and mobile clusters of small households, avoiding contact with each other and non-Lacandons. According to Paladino (2005, p. 116), until the 1930s, the Lacandons had mainly been in touch with lumbermen and *chicleros* in their camps as outside contacts, with whom they sometimes traded tobacco and forest products for manufactured goods and salt but whose diseases they sought to avoid. This is why the Lacandons purposefully allowed the trails to their own milpas and houses to disappear in the forest to obscure their destinations.

Paladino (2005) cites, as her sources for this information, the dissertation of James Nations, who spent four decades in the region as an anthropologist and conservationist and has actively promoted the Maya Forest concept (Nations, 2006). At present, while writing these lines, I do not have access to his early works from the 1970s. However, I am mentioning this detail here as, in his new book (Nations, 2023, p. 12), he seems to conclude something different entirely: Instead of *chicleros,* he indicates – almost with the same sentence as above – that the Lacandons traded with the *missionaries*. Instead, he refers to the *chicleros* as follows:

> Survival in the Selva Lacandona is a full-time task, even for people who have lived there all their life, and the outsiders, who sought out the Lacandones seemed intent only on stealing their land, felling their trees, or changing their religion. Not to mention the crocodile hunters, chicle gum

harvesters, and all-purpose renegades who periodically raided family compounds to abduct women and girls.

(Nations, 2023, p. 11)

This is a surprising argument. I wonder if the following mention of *chicleros* a few pages later explains such a comment: "Through history, some Lacandon women have left their families to marry non-Lacandon men, most frequently chicle harvesters or mule drivers, but many of them later returned home with mixed-race children and without their husbands" (Nations, 2023, p. 17). Finally, in his Lacandon rainforest timeline, he mentions that in the 1940s, "chicle gum harvesters roam the rainforest to tap chicle trees for chewing gum base, interacting with Lacandones in both the Guatemalan Petén and Selva Lacandona" (Nations, 2023, p. 252).

Whatever the relationship between *chicleros* and Lacandones may have been, the Mexican Commission of Protected Areas (CONANP, 2006) also suggests that the *chicleros*, together with lumber laborers and oil explorers, opened trails and established connectivity between the Lacandons and others. With the exodus from the 1930s onward, the Lacandons began withdrawing deeper into the center of the forest.

As a result, new processes ensued: The creation of a cultural mosaic that shifted the earlier *finca*-based (estate) system toward *ejidos* (communal lands) that contributed to increasing struggles of land rights, and the saturation of natural resources and complex organizational processes, involving political-religious and peasant struggles related to the deepening agrarian conflict, among others (Leyva Solano & Ascencio, 2002). As is well known, the catalyst for the agrarian conflict was the 1971 presidential decree (1972 resolution) that granted the newly formed Lacandon Community (which included 66 Lacandon families) 614,321 hectares of the rainforest while leaving out 26 Indigenous settlements already established in the same lands (Leyva Solano & Ascencio, 2002). According to De Vos (2015), the settled colonies were more than 40, out of which several had already been formalized as *ejidos*, while others had been in the process but had been suddenly declared illegal by the decree. Several private landowners, *latifundistas*, also lost their lands in one strike.

As extensively narrated by De Vos (2015), the presidential decree, elaborated with some haste, contained remarkable mapmaking errors to the degree that it could be described as a cartographic illusion corresponding to a political forest. By cartographic illusion, Ingold (2000, p. 234) referred to a mapmaking that suppresses or brackets out both the movement of people as they come and go between places and the re-enactment of those movements in inscriptive gesture (mapping), thereby creating an illusion based on an assumption about the world that the map was supposed to represent – like a theatrical stage, from which all the actors have mysteriously disappeared. In this case, full communities were mysteriously neglected. The designated hectares did not coincide with the totality of the designated area either. In addition, the decree created something called the "Lacandon zone", although the actual Lacandons were settled in several distant places, some of which were left out of the map. Moreover, according to the decree, the Lacandon zone included such a place as *Zapote Caribal*, which apparently never existed. While the decree also announced hectares for "conservation of archaeological sites" and "ecological conservation", in reality, the named ruins were mapped as part of the Lacandon community, while the protected lagunes remained completely outside the decree.

In the account of De Vos (2015), the decree was influenced by the archeologist couple Gertrude Duby and Frans Blom, who, since the 1950s, had explored the Mayan ruins of the Lacandon rainforest while collaborating with the group of Lacandons living there. During the stays and explorations with the Lacandons, the power couple, Duby and Blom, witnessed the accelerated process of colonization and deforestation and were in a hurry to protect what they considered the lands of the Lacandons. According to Trench (2008), Duby had a considerable effect on the public image of the forest-dwelling Lacandons as the authentic, original Mayas who then became objects of keen anthropological interest and international tourism. Both De Vos (2015) and Trench (2008) note that the hasty decree, in addition to the urgency to conserve the rainforest, was influenced by a pact with the Lacandons to allow logging, which took place during the years after the decree.

However, from the 1970s onward, the Lacandon Community attempted to take control of the forestry and to eliminate intermediaries that had benefited from the timber trade. Among others, Trench and Köhler (2004), Cano (2018), and Paladino (2005) extensively narrate the arduous forestry history of the Lacandon Community entangled with the agrarian conflict. As part of the history, the Lacandon Community intended to organize *chicle* extraction around the Usumacinta River, which had, until then, pertained to outsiders concessioned by the federal government and who were, in the politized process, even taken as hostages. The extraction of *chicle*, however, became unprofitable in the beginning of the 1980s, and thus, the cooperative then focused on *xate* (Trench & Köhler, 2004). Toward the 1980s, the conflicts related to timber and non-timber extraction increased and, in many ways, culminated in the forest ban (*veda forestal*) between the years 1989 and 1994 (e.g., Cano, 2018). During this time, the sociopolitical situation radicalized, and when the peasants discovered governmental timber extraction despite the ban, revolts ensued. In the process, for example, an organization called Independent Revolutionary Peasant Movement (MOCRI) was established, which was dedicated to mobilizing a broad social base that allowed the peasants to take the forestry in their own hands (interview with a MOCRI-founder in Marqués de Comillas, Chiapas, January 2019). Our interviews in 12 different *ejidos* in Marqués de Comillas in 2019 suggested that MOCRI was considered one of the main actors in the region, fighting to position the forests in the hands of the communities in a similar vein as the forest concessioners in the Maya Biosphere Reserve. The other important actors included, of course, the conservation organizations. Both instances were alluded to as having represented "substantial interventions" in the recent history of the rainforest's southeastern parts.

The radicalization was also linked to various other simultaneous processes in the southeastern areas of the Lacandon rainforest along the Lacantún and Usumacinta Rivers. In the 1970s and 1980s, posterior to the oil explorations in the area, and when the Guatemalan civil war deepened, the Mexican state became highly consumed with fostering the international border by declaring the southern Lacandon rainforest "open land for colonization" and, in particular, for cattle ranching. Thus, in the footpaths of oil explorers, the lands bordering the Lacantún River, next to the Montes Azules Biosphere Reserve near the Guatemalan border, were settled. In the 1980s, these areas experienced heightened deforestation, forest fires, illegal trafficking, and Guatemalan refugees. The situation culminated more broadly in 1994, with the Zapatista insurrection. As a result, in 1998, the southeastern parts of the rainforest, until then pertaining to the grand Ocosingo-municipality, were administratively divided and reformed as new, smaller municipalities – in this case, Marqués de Comillas.

These developments have been intimately connected to nature conservation. The Montes Azules Biosphere Reserve that covers most of the remaining Lacandon rainforest was created in 1978 with the pressure exercised by the previously mentioned Gertrude Duby, among others. According to Trench (2008), this took place without prior consultation with the forest inhabitants, including the Lacandon Community that was supposed to be – and was, to varying degrees, recognized as – the guardian of the reserve lands. The biosphere reserve was later followed by other protected areas and archaeological sites in the region, particularly near the Lacandon Community lands.

The conservation organizations involved in the reserve became both targets and go-betweens in the regional conflict of the 1990s as, while the ongoing politicized and militarized dynamics unfolded, the reserve needed to set its management plan. The management planning process evidently coincided with the other simultaneous governmental interventions. Paladino (2005) exhaustively narrates the context of the 1990s characterized by the environmental "hardcore" governmental approach that concurred with attempts to control the political unrest with militarization, combined with other risky situations, such as paramilitary activity and deforestation in the forest space, which took place due to the earlier governmental policies encouraging colonization and cattle ranching. In other words, environmental discourse and legislation that contradicted the earlier policies, while deepening the overall politicization newly emerged. My interviewee from OMIECH commented that, in the 1990s, "the forest was a closed space" and that most communities, particularly sympathetic to the Zapatistas, refused any government-linked projects, including those of conservation, as they represented counterinsurgency. Indeed, as Paladino (2005) notes, conservationists were criticized for misled funding, lack of transparency, foreign involvement, commercialization of species, and the involvement of multinational companies, among others. The Lacandon Community itself was particularly critical about their role as rangers and guardians in charge of the forest and in receiving funds for those functions.

The conservationists evaluating the situation corresponding to Montes Azules Biosphere Reserve in the 1990s concluded that to address and deflect the "pressure points" on the reserve's peripheries, a new strategy of *stations* could be introduced (Paladino, 2005, p. 207). The conservationists running the stations consider these "strategic spaces" that allow creating connectivity; in other words, places-in-knots between different actors related to reserves, such as local people, authorities, technicians, scholars, and visitors. They may also allow local employment possibilities. In the 1990s, when the strategy was implemented, two stations already existed in the southeastern areas of the reserve, one of which had been established in the early 1980s and had been used for ecological research. This was located in Marqués de Comillas, along the Lacantún River in close proximity to the Guatemalan border, and in the southern periphery of the reserve pertaining to the Lacandon Community. The other stations were planned for ecotourism, environmental education, and research closer to the Lacandon Community. According to our group interview conducted in the installations of the research station in January 2019, the origins of the station were part of a broader process related to the creation of similar stations in various protected areas in Mexico. In this case, the lands of the station were rented from the Lacandons, who had agreed that the southern parts of the reserve would benefit from the scientists' presence and vigilance. According to Paladino (2005), the Lacandon Community agreed to rent in exchange for clarity with regard to the funds and projects and for their community members to be hired as guards in the reserve. The station does not really patrol but rather monitors the area. The patrolling task pertains to the previously mentioned CONANP.

Image 5.8 Lacantún River and the Montes Azules Biosphere Reserve, Mexico
Source: Hanna Laako, 2020

According to Berget (2021, see also Carabias et al., 2015), who interviewed the nearby *ejido* regarding its history, the station's site was first visited by biologists in 1978, but it was abandoned due to lack of resources and even vandalized during the 1980s. Between 1982 and 1984, the nearby *ejido* received approximately 5,000 Guatemalan refugees that were later transferred to Campeche and Quintana Roo; however, during this time, the area was deforested for subsistence needs. From 1989 onward, the station was revived, and in 1992, it received funds to build the installations.

The conservation organization currently managing the station was established in 2005 by leading Mexican conservationists and biologists to conserve the Mexican southern tropical rainforests. The founders were the same involved with the station from its beginning. While the organization has focused on biodiversity monitoring, it has also engaged in a considerable effort to build communitarian ecotourism, environmental education, and territorial reordering in five surrounding communities on the other side of the Lacantún River in the municipality of Marqués de Comillas. It is also involved in transboundary efforts, such as the conservation of the jaguar and red macaw. Additionally, the organization also works on improving environmental legislation and conducting scientific projects.

During the past decades, the Lacandon rainforest has lost more than half of its original coverage and has succumbed in many violent socioenvironmental processes that include extensive cattle ranching, African palm oil plantations, wildlife smuggling, looting, and some sparse drug landing strips, among others. In this context, in 2014, the leader of the conservation organization was abducted from the station. Accompanied by students in the station, the group was attacked by men, who apparently took (only) the leader by boat and foot to the Guatemalan side, which is in close proximity, and from where the complex negotiations took place. Eventually, the leader was released. She later commented in an interview that as the organization is defending "the last bastions" of the rainforest, and is opposed to further concessions within the reserve, its enemies have grown (Rabasa, 2019). Environmental activism in Mexico is one of the most dangerous endeavors after journalism.

The conservationists in our interview noted an increase in the number of settlements within the reserve. However, the problem was difficult to address given the agrarian conflict. The term "illegal settlement", for example, is politicized given that many local people argue that they cannot be tagged as illegal if they were there prior to the reserve or have settled there as a result of the problems generated by the agrarian conflict deriving from earlier failed state policies. The ongoing discussions about the settlements inside the reserve also provoke complex fears: Whether these local people are collocated elsewhere with new land entitlements or allowed to stay, this could signal a reward for further "invasions". Our interviewees at the station also commented that many conflict situations related to smuggling, money laundering, and cattle ranching are easily redirected as those of "Guatemalans". While the organization has a strong profile in conservation biology and the station is strictly focused on scientific activities, they have considered it necessary to engage in socioeconomic support and projects with the communities as, "otherwise, the forest is gone".

Indeed, no tourism was allowed at the station when we visited it in January 2019. Instead, the tourist lodges and services are provided by the nearby communities as the station only receives its members and researchers, mainly biologists. Given the tense situation, even these visits are sometimes canceled and planned under strict safety measures. Our arrival instructions included finding a particular house in a particular *ejido* and a man with a pseudonym, who called by radio to the station so that the guards – also community members – could pick us up from the shore, cross the river by boat, and enter the station. During our days there, when we mostly visited the area for communal interviews, we also observed the biologists in their work environment, with early-morning preparations to dive samples at the river, late-night insect traps, and red macaw feeding, accompanied by the nearby sounds of howler monkeys, in addition to a trail through the forest to learn about the trees and species, including the sapodilla. As with the other stations, this one is also filled with footpaths toward the interior of the reserve, and these trails are a frequent mention of the researcher's exploring biodiversity in the surroundings. One of them is even said to lead to *El Zapote* ruins, a vague archaeological site located within the reserve, the name of which could derive from the sapodilla tree fruit also pictured by the ancient Mayas in Palenque (Mathews, 2009). Whose trails these originally are, and if they contain scarred *chicle* trees, remains an open question that I was not able to resolve. Thus, my *chicle* route ends right here where the sapodilla, again, fades away in the shadows of a political forest.

Image 5.9 Forest Trail in the Biological Research Station in the Montes Azules Biosphere Reserve, Mexico
Source: Hanna Laako, 2019

Conclusions: In the Shadows of a Sapodilla Tree

A sapodilla tree can live up to 100 years and grow more than 30 meters high. The sapodilla trees of the Maya Forest – particularly the scarred ones – may have witnessed the dramatic historical epochs unfolding in their shadows. The sapodilla tree is well equipped for harsh conditions. It has adapted to karst limestone regions but grows also in sandy places. It is highly resistant to drought, heat, and hurricanes and is known for its extreme longevity and resilience to disturbance (Mathews, 2009).

The sapodilla tree has also developed a particular healing power: if attacked by insects or cut by humans, the tree produces a milky fluid that forms a protective layer over the damaged area. This is the latex that has been used for centuries in Mesoamerica, particularly by the Mayas, who used the *chicle* in numerous ways. The ancient Mayas referred to the tree as "*tzicte'ya*", which roughly translates to "wounded noble tree". Indeed, according to Mathews (2009), the tree was so valuable that depictions of the sapodilla can be seen on ancient Maya sarcophagus – for example, in Palenque, Chiapas.

Thus, it does not seem to be totally a coincidence, nor completely dependent on waterholes, that the *chicle* hunters of more contemporary era have found the best sapodilla trees in the surroundings of ancient ruins. According to Mathews (2009), several authors have paid attention to the high concentration of sapodillas around archaeological sites. The Mayas seem to have planted the valuable sapodilla trees in their forest gardens. However, another reason might have been related to a biocultural symbiosis: the most important pollinator of the sapodilla trees are *bats*, and again, the archaeological sites in Mesoamerica have an abundance of bats and sapodilla trees. A prime example is the ruins of Calakmul, which are in close proximity to one of the world's largest bat caves. Some researchers have suggested that the wide distribution of sapodillas in Mesoamerica is not limited to human activity but also includes that of bats.

Thus, a curious biocultural chain of Maya Forest history emerges, connecting the sapodillas, bats, the Mayas, the *chiclero*, the settlements, archaeologists, and conservationists. Perhaps the sapodilla tree is a bit like the Maya Forest itself: scarred but resilient in its place, with a healing power, and a bit in the shadows of the more famous ones, such as the ceiba, Amazon, Lacandons, and the Morley.

This chapter has revisited some of the roots and routes of the sapodilla tree, which, during the research running from 2019 to 2024, made a constant appearance – even disturbance – until I started to pay attention: the formative but somehow shadowy historical role of the *chiclería* in the contemporary Maya Forest. The archaeological and conservationist trails turned out to be those of *chicleros*, and the stations to be those of *chicle* camps. While the existing scholarship on Maya Forest *chiclería* has focused on certain histories often trapped by national borders, the stories from the trails suggest mobility and connectivity within the Maya Forest that contributed to creating new places-in-knots. Some of these places-in-knots are transboundary, such as the

Chiquibul or Laguna del Tigre. Others are settlements, communities and *ejidos* established as a result of the *chiclería*, such as those within the Lacandon, Maya Biosphere Reserve, and Quintana Roo. Thus, the *chiclería* does not represent only foreign influence but an inner regional connectivity in supposedly remote places. As such, it has also contributed to the ecological knowledge accumulated within those regions and places.

Evidently, the *chicle* trails also reveal the role of the oft-unnamed go-betweens, the *chicleros*, and the global power relations that include science and Archaeology. The example of *chiclería* clarifies the challenges related to such a complex as the Maya Forest, entangled in powerful scientific and Mayanist ambitions and projections. Their trails help to unfold the layers of the political forests, whereby *chicleros* created alliances both as physical and transactional go-betweens with another type of representational go-betweens, those of archaeologists, which now entangle with new generations of actors, such as forest concessionaries and forest peasants.

Political forests, as originally defined by Peluso and Vandergeest (2011), address nation-states and governments as mapmakers that may inflict a great deal of structural violence and conflict with their cartographic illusions. The *chiclería* is a prime example of cartographic amnesia, whereby governments forget about settlements, migrations, and histories that are the results of their own previous policies to employ new, supposedly better ones.

Yet, apart from what such an enterprise can cause to lives, human, and/or wildlife, we are all also guilty of a mapmaking amnesia to some extent, bound as we are in our own partial lenses through which we try to gaze at the realities and routes.

Chicle history is also about a particular angle on capitalism: what it leaves behind. I wonder if many critical studies related to capitalism also tend to forget about *the things left underfoot*, as Borderlands Studies emphasized in Chapter 1, when extractive businesses find their new resources and address their attention to other areas to be conquered and exploited, and which are then followed by the researchers' keen, elucidative eye. What was left behind – the other kinds of *ruins* – are far too easily left to invisibility, and a double-shame: shame for having followed the business and shame for having been left behind as good-for-nothings, now without subsistence, at least until the capitalist gaze finds their places again but perhaps with a new plan in which they do not fit– new names, new mapmaking. Yet, these are places conformed by the experience, history, lives, families, and sense of community that generate pride among the people: the places-in-knots, the wayfinding. This is also what I sensed in my interviews related to *chiclería*, despite the harsh conditions described. Thus, I can see *chicle* history not only through a lens of misery but through one of agency.

While the conservation mapmaking may prioritize the ecosystems, within their trails and stations, they are immersed in socioenvironmental and political *places*, engaged in political forests as actors and mediators. As a researcher (yet another kind of go-between), my own trail in the Maya Forest waterlands' places-in-knots, such as the biological stations, allowed me to understand more complex regional and transboundary dynamics that I would have missed by looking at mere maps. In these places, many responsibilities seem to have been outsourced by the governments to local actors and nongovernmental organizations, while the governments remain within their right to intervene, control, or retire to amnesia.

In this sense, I have come to think of a concept of political forest not solely reduced to national boundaries or governmental cartographic illusions as a close-ended space, which eventually extracts the political, that is, the place, histories and hopes, out of the picture. While this sort of mapmaking is also part of the Maya Forest waterlands' reality, as shown in other chapters of the book, political forests need to address the places-in-knots, trails, and go-betweens, which are often those in the middle of cartographic mapmaking and wayfinding.

Notes

1 *Paraje* refers to a remote place or location in Spanish.
2 *Ladino* is often used as a synonym for *mestizo*, that is, mixed Spanish and Indigenous heritage, usually Spanish speaking by birth. Sometimes, it can also refer to Spanish-speaking Indigenous depending on assignment criteria (language, clothing, or self-ascription).
3 *Milpero* meaning the one who has a milpa, a small agroforestry (predominantly corn) field in Mexico or Mesoamerica.
4 *Co-management* in conservation refers to shared duties and rights within protected areas. These are often governmental protected areas, which are mostly managed by different kinds of nongovernmental organizations.
5 *Peace parks* refers to a system of transboundary protected areas created to enhance peacebuilding in conflict-prone borders (e.g., Ali, 2007).
6 *Aguardiente* refers to alcohol spirit.

References

Ali, S. (Ed. 2007). *Peace parks: Conservation and conflict resolution.* MIT Press.

Awe, E. (2004). Chicle extraction and its operation in Old El Cayo. *Nostalgia: The Voice of Experience* March/April, 5–9. Octavia Waight Centre Publication.

Berganza, K. (2017). *The Impact of the Chicle Industry in Western Belize: Social and economic aspects of the effects of the Chicle industry in Western Belize.* University of Belize. https://www.academia.edu/35362533/The_Chicle_Industry_in_Western_Belize_socio_economic_effects_of_the_Chicle_Industry_in_Western_Belize.

Berget, C. (2021). *Un recorrido por la historia de Boca de Chajul: Como se formó, sus habitantes y su paisaje.* Wageningen, ECOSUR et al. https://www.researchgate.net/publication/361138415_Un_recorrido_por_la_historia_de_Boca_de_Chajul_como_se_formo_sus_habitantes_y_su_paisaje.

Cano, I. (2018). *De montaña a "reserva forestal": Colonización, sentido de la comunidad y conservación en la selva Lacandona.* UNAM.

Carabias, J., De la Maza, J. & Cadena, R. (Eds.) (2015). *Conservación y desarrollo sustentable en la Selva Lacandona.* Natura Mexicana.

CONAP (2019). *Plan maestro de los complejos III y IV: Áreas protegidas del sureste de Petén.* Asociación Balam y FONACON.

CONANP (2006). *Programa de Conservación y Manejo: Área de Protección de Flora y Fauna Nahá.* CONANP.

De Vos, J. (2015). *Oro verde: La conquista de la Selva Lacandona por los madereros tabasqueños 1822–1949.* Gobierno del Estado de Tabasco and Fondo de Cultura Económica.

De Vos, J. (2015). *Una tierra para sembrar sueños: Historia reciente de la Selva Lacandona 1950–2000.* CIESAS and Fondo de Cultura Económica.

Devine, J. (2016). Contesting Global Heritage in the Chicle Workers' Museum. *Latin American Research Review* 51(3), 101–122.

Devine, J. A., Wrathall, D., Currit, N., Tellman, B. & Langarica, Y. R. (2020). Narco-Cattle Ranching in Political Forests. *Antipode* 52, 1018–1038. doi:10.1111/anti.12469

Duffy, R. (2022). *Security and Conservation: The Politics of the Illegal Wildlife Trade.* Yale University Press.

Dugelby, B. (1998). Governmental and customary arrangements guiding Chicle latex extraction in the Petén, Guatemala. In R. Primack, D. Bray, H. Galletti, & I. Ponciano (Eds). *Timber, tourists and temples: Conservation and development in the Maya Forest of Belize, Guatemala and Mexico* (pp. 155–179). Island Press.

Fiscella, B. (2020). *Memories of the future: The modern discovery of a relict ecosystem.* Film documentary. https://vimeo.com/493503303.

Forero, O & Redclift, M. (2006). The role of the Mexican state in development of chicle extraction in Yucatán and the continuing importance of coyotaje. *Journal of Latin American Studies*, 38, 65–93.

Ingold, T. (2000). *The Perception of the environment: Essays on livelihood, dwelling and skill.* Routledge.

Kauffer, E. (2023). Negotiating smallness in three regional contexts: Belize within Central America, the Caribbean and neighboring Mexico. In T. Kolnberger & H. Koff (Eds). *Agency, Security and Governance of Small States: A Global Perspective* (pp. 104–118). Routledge.

Kauffer, E. (2021). De las selvas de las fronteras a las fronteras de las selvas: Una historia de ríos y de fronteras entrelazadas. Reseña. *Península* XVI(1), 245–249.

Konrad, H. (1987). Capitalismo y el trabajo en los bosques de las tierras bajas tropicales mexicanas: El caso de la industría chicle. *Historia mexicana* 36(3), 465–505.

Legorreta, C. (2008). *Desafíos de la emancipación indígena: Organización señorial y modernización en Ocosingo, Chiapas 1930–1994.* UNAM.

Leyva Solano, X., & Ascencio, F. (2002). *Lacandonia al filo del agua.* CIESAS, UNAM and Fondo de Cultura Económica.

Lowenhaupt Tsing, A. (2015). *The mushroom at the end of the world: On the possibility of life in capitalist ruins.* Princeton University Press.

Martínez-Reyes, J. (2016). *Moral ecology of a forest: The nature industry and Maya post-conservation.* University of Arizona Press.

Mathews, J. (2009). *Chicle: The chewing gum of the Americas, from the ancient Maya to William Wringley.* The University of Arizona Press.

Matola, S. & Platt, E. (1998). One forest, two nations: The Chiquibul Forest of Belize and Guatemala. In Primack, R., Bray, D., Galletti, H. & Ponciano, I (Eds). *Timber, Tourists and Temples: Conservation and Development in the Maya Forest of Belize, Guatemala and Mexico* (pp. 23–29). Island Press.

Metcalf, A. (2005). *Go-betweens and the colonization of Brazil 1500–1600*. University of Texas Press.

Millner, N., Peñagaricano, I., Fernandez, M. & Snook, L. (2019). The politics of participation: Negotiating relationships through community forestry in the Maya Biosphere Reserve, Guatemala. *World Development* 127, 104743.

Morley, S. (1938). *Inscriptions of Petén*. Volume 1. Carnegie Institution of Washington. https://hdl.handle.net/2027/uva.x001476617

Nations, J. (2006). *The Maya tropical forest: People, parks and ancient cities*. University of Texas Press.

Nations, J. (2023). *Lacandón Maya in the twenty-first century: Indigenous knowledge and conservation in Mexico's tropical rainforest*. University Press of Florida.

Paladino, S. (2005). *We are the guardians of the selva: Conservation, Indigenous communities and common property in the Selva Lacandona, Mexico*. Doctoral dissertation. University of Georgia.

Peluso, N., & Vandergeest, P. (2011). Political ecologies of war and forests: Counterinsurgencies and the making of national natures. *Annals of the Association of American Geographers* 101(3), 587–608.

Pérez Aguilar, R. (2014). El chicle en Quintana Roo: sus caminos y voces. *Cuicuilco*, 21 (60), 195–222.

Pérez Mendoza, J. (2023). "Este olvidado pedazo de nuestra patria": los proyectos del gobierno mexicano para controlar la franja fronteriza de la Selva Lacandona y aprovechar sus recursos naturales, 1950–1970. *Revista Pueblos y fronteras digital*, 18, 1–33.

Ponce, M. P. (1990). *La montaña chiclera Campeche: Vida cotidiana y trabajo (1900–1950)*. CIESAS.

Prado, F. (2012). The fringes of empire: Recent scholarship on colonial frontiers and borderlands in Latin America. *History Compass* 10(4), 318–333.

Rabasa, D. (2019). La defensora de la Selva. *Gatopardo: Reportajes*. https://gatopardo.com/reportajes/julia-carabias-una-de-las-mayores-guardianas-del-medio-ambiente-en-mexico/

Rahder, M. (2020). *An ecology of knowledges: Fear, love and technoscience in Guatemalan Forest conservation*. Duke University Press.

Ramsar (2024). *Ramsar site information service: Parque Nacional Laguna del Tigre*. https://rsis.ramsar.org/ris/488?language=en

Rice, P. & Ward, C. (2021). *The archaeological field diaries of Sylvanus Griswold Morley 1914–1916*. www.mesoweb.com/publications/Morley/Morley_Diaries_1914-1916.pdf

Rodas, I. (2014). Las colonizaciones y la frontera nacional en la selva petenera, Guatemala. *Cuadernos CEMCA: Ciencias Económicas y Sociales* 7, 7–10.

Roessingh, C. H. & Darwish, K. (2012). Self-employment and chicle trade: The case of Lebanese minority in the Cayo District of Belize. *International Journal of Entrepreneurship and Small Business* 17:1, 1–16.

Saxer, M. (2022). *Places in knots: Remoteness and connectivity in the Himalayas and beyond*. Cornell University Press.

Schwartz, N. (1990). *Forest society: A social history of Petén, Guatemala*. University of Pennsylvania Press.

Smouts, M. C. (2003). *Tropical forests, international jungle: The underside of global ecopolitics*. Palgrave Macmillan.

Sundberg, J. (2003). Conservation and democratization: constituting citizenship in the Maya Biosphere Reserve, Guatemala. *Political Geography* 22:7, 715–740.

Taylor, P. L. (2010). Conservation, community and culture? New organizational challenges of community forest concessions in the Maya Biosphere Reserve of Guatemala. *Journal of Rural Studies* 26, 173–184.

Torras, R. (2019). *Colonización y colonialidad en una selva de frontera. La cuenca campechana del río Candelaria (siglos XIX y XX).* UNAM.

Trench, T. (2008). From "orphans of the state" to the *Comunidad Conservacionista Institucional*: The case of the Lacandón Community, Chiapas. *Identities: Global Studies in Culture and Power* 15(5), 607–634.

Trench, T. & Köhler, A. (2004). Los xateros en la selva Lacandona: Una temporalidad permanente. *Anuario 2004*, 351–382.

Tuck, E. & McKenzie, M. (2015). *Place in research: Theory, methodology and methods.* Routledge.

Tzul, A. (2004). The development of the chicle industry in El Cayo. *Nostalgia: The Voice of Experience*, May/June, 4–5. Octavia Waight Centre.

Vera-Martínez, P. & Ceballos-Falcón, E. (2024). Women's empowerment and the honey production projects in the protected areas of Usumacinta Canyon, Mexico. *Journal of International Women's Studies* 26(1), 10.

Ybarra, M. (2018). *Green wars: Conservation and decolonization in the Maya Forest.* University of California Press.

The Maya Forest Waterlands, or There and Back Again

Hanna Laako and Edith Kauffer

It was the first day of our initial exploratory fieldtrip for a megaproject related to the transboundary Usumacinta River basin. On a cold mid-January morning of 2018, we had just been seated in two boats at the edges of the Usumacinta River in Tenosique, Tabasco. Our group included natural and social scientists, a representative of the Mexican Civil Protection Agency, and local fishermen. Our objective was to travel upstream the Usumacinta River, observing this historical boundary, access place, and watery resource while camping for several nights on its different shores, and eventually arriving at the town of Frontera Corozal in Chiapas. Our social science team had left a day before from Chiapas, traveling through Villahermosa, Tabasco, picking up one of the boats from the offices of the Civil Protection Agency, and continuing our way toward Tenosique, together with the natural science team that had joined us from Villahermosa. The local fishermen, who have lifelong experience in navigating the river, had agreed to act as our guides and boat drivers.

Early this morning, we were ready to start our journey from Tenosique, divided in two boats. Our belongings were stuffed in black plastic bags, and our guides gave us last-minute instructions about leaving our shoelaces untied in case of contact with the water, while ensuring our life jackets were in place. It had been raining heavily in the Guatemalan highlands days before, and the Usumacinta River was effervescent. However, the sun was coming up, and we spent the following couple of hours on the boats, observing and taking pictures of human activities and natural features, while our boats passed by these "Maya Canyon" Tabasco-Chiapas borderlands.

However, our trip was going to be short lived. Soon, we arrived at the outskirts of the Great San José Rapids. A bit closer to the rapids, the fishermen stopped the boats. Making last-minute plans, they decided that each boat would take its turn to cross the rapids in a *z*-like maneuver. The boat where Kauffer was sitting was the first in turn, while the second boat, where Laako was located, remained behind, waiting. From the waiting boat, Laako with her companions observed the first boat ahead in the distance, proceeding in slow motion toward the rapids and then through the foaming torrent. Everybody was thrilled, taking pictures.

DOI: 10.4324/9781003429050-7

Suddenly, however, the first boat seemed to stagnate in the rapids, and the fishermen on Laako's boat started to shout instructions to "give it the full motor!". Yet the boat did not seem to progress. On the contrary, to our dismay, the boat vanished in the blink of an eye. Then, we saw the crew and luggage circling in the water, drifting fast toward us in the powerful stream. There was short-lived, complete silence on Laako's boat. Cameras were laid down. In the next confusing moments, hands extended to reach our companions in the water, and instructions were shared. Those who could – among others, Kauffer – swam to the shore, while others attached to Laako's boat as it headed toward the same rough Usumacinta edges.

After some chaotic moments, when everybody had reached the rocky riverside, the fishermen took the surviving boat to save the shipwrecked one, which was still whirling upside down in the rapids. Many belongings, such as an expensive drone, were picked up from the angry waters thanks to the well-tied plastic bags that allowed them to float, whereas some not-so-fortunate luggage was lost to the depths of the river, such as Laako's backpack. Everybody was sitting on the rocks, trying to rescue what could be rescued – wet cameras, clothes, and other equipment.

After some pondering, everybody agreed that the only reasonable plan was to return to Tenosique. Some hours later, another boat came for us from Tenosique and took us all back to the town. The next day, and after some emergency shopping, we continued the journey by car, visiting the "forest waterlands" of the river in several different locations along the Mexican–Guatemalan border.

Image 6.1 The Boat Carrying Edith Kauffer in the San José Rapids of the Usumacinta Canyon a Few Seconds Before it Disappeared under Water (Fieldwork in January 2018)
Source: Hanna Laako, 2018

It was later confirmed that our shipwrecking, which by now has become "the ultimate, legendary fieldwork story" among colleagues in southern Mexico, had been caused by a motor failure, aggravated by the stream, which had grown heavier due to the earlier rains in the Guatemalan highlands. The Usumacinta River had shown us its force on one strike, effectuated by its transboundary nature, which took us quite literally to its watered lands.

This was surely an introductory plunge in the waters of Usumacinta, which also gave us a very concrete, empirical touch to its fluid borders, entangled politics, and shared conservation. Thus, it could be said that this experience was a starting point for the topics that now culminate in the pages of this book as the Maya Forest waterlands. *The Maya Forest Waterlands* is indeed an invitation to critically examine the interstices, edges, and nexus of the Maya, forests, waters, and lands where different people and wildlife interact, as well as to shape and challenge contemporary borders and politics.

The Maya Forest waterlands represent, among other things, a strong call to examine the practice of science: The developments of various fields from their early explorers and adventurers to today's globally established disciplines, including the power of mapping and naming, the (de)coloniality of knowledge, as well as the different power relations embedded in academic practices in given locations and positionalities, and between countries, institutions, and scholars. It has incited us to examine many epistemological, theoretical, and methodological foundations, collaborations, and borderings. In a way, it has implicated a two-way analysis of the ways in which the region's processes and phenomena – whether human, natural, or both – modify and challenge these scientific histories and narratives and the ways in which they shape the region's dynamics and storytelling. Interchangeably, they cast lights and shadows on different issues. In this sense, it is a coming-and-going, there and back again, as the subtitle of this afterword indicates, borrowed from J.R.R. Tolkien's *The Hobbit*.

As the chapters of this book have shown, the Maya Forest is a construct that reflects some major scientific developments and impacts, such as the creation of the Maya and of a biodiversity hotspot with its ecoregions, which do not only illustrate the region today but take part in shaping it. This is so much so that throughout the chapters of this book, we have pinpointed that the transboundary region can now hardly be considered a periphery, but rather should be considered an eco-borderlands that involves many human-nature entanglements, as well as substantial territorial, and bio- and geopolitical transformations. Those are, among others, tourism entailing gentrification, urbanization, and new infrastructure; drug-cartel activity entailing violence and militarization; conservation entailing territorial reordering and land use; and many others – some governmentally instigated and others related to peoples' necessities. Some of these developments are more extractive, while others seek to be curative. Many involve the two in an entangled manner.

The study of these entanglements is important because it allows us to understand certain challenges – and perhaps also potential solutions. Most importantly, the examination of the nexus and interstices shows that certain processes and phenomena do not exist in a vacuum in an interconnected world. In times of climate change and biodiversity loss, the study of connectivities has become even more crucial and difficult. Our tools as humans are limited, and thus we create simplified categories and draw borders to administer the immediate world as we see fit.

This has not been a book about public policy recommendations though – as social scientists, we are focused on critical analysis destined to shed light on complex realities. We might even ask the critical question of whether public policies continue to use the most efficient tool to address these contemporary challenges. In many ways, the Maya Forest itself is a creation to reach beyond national boundaries and policies and to enable a space for shared conservation work, past particular government agendas and methodological nationalism.

We have stretched this further with the Maya Forest waterlands, arguing that this challenging analytical choice has been paramount in carving out certain important broader tendencies that otherwise could have seemed to be isolated cases. For example, the broad fieldwork visiting different biological stations and studying their origins and backgrounds brought up the unexpected, shared regional history of *chiclería.* We maintain that this is a key phenomenon for understanding the Maya Forest waterlands. During the colonial era, the region had suffered an extreme decline in human population, which nevertheless resulted in a rewilding secondary forest. This only began to change at the end of the nineteenth century and, particularly, with the *chicle* boom that mobilized peoples across the region toward the depths of the forest waterlands. This resulted in the formation of new localities at the same time with the formation of national borders, much later followed by the borders of protected areas. Yet many of these communities remain. These communities and trails were intimately connected to the Mayan ruins and rediscovered by these people. Some of them were local and mobile, while others were newcomers and settlers, who were encouraged by the emerging nation-states, their businesses, and borderings. These were the emerging postcolonial forest people that are the actors and inhabitants of the Maya Forest today. They are both Mayas and non-Mayas. And yet, so far, they have been the least visible in the predominant Maya Forest narratives.

As a result, two interconnected aspects ensue. On one hand, the introduction of the Maya Forest concept, with its invitation to broader, regional, and transboundary analysis, has enabled us to realize the extension and importance of this mentioned *chicle* history, materialized in an important symbiosis between the Maya/forest/water/lands and its borders in the making. On the other hand, it challenges the predominant Maya Forest narratives centered around the ancient Mayas as the only actors, discovered by archaeologists and safeguarded by conservationists in the protected areas along national borders. According to *chicle* history, that of the Maya is not the only shared heritage of the region.

These challenge us also to rethink rurality and remoteness. While the Maya Forest waterlands may still be perceived as a predominantly rural landscape, distant from its national centers (at least in the case of Guatemala and Mexico), it invites us to ponder: In which sense is it remote? Policies deriving from political power centers are often addressed and implemented in their edges, purposefully defined as resourceful yet "empty" or remote lands ripe for new developmental agendas. This is often combined with amnesia toward previous programs that affected and mobilized peoples' lives in the forest waterlands. Thus, borderlands are also places of accumulation of state interventions, old and new, many of which are later forgotten, rusting away or cast aside from the way of new aspirations. In this way, borderlands are locations for different cycles of rediscovery and possession. This might be the case particularly with those borderlands located on international borders, such as the Maya Forest, where governments always hold the potential geopolitical interest to enhance their presence.

In this way, rurality and remoteness may work as political tools, while consisting of real living conditions, characterized by limited infrastructure, services, and connectivity with the national or global centers. As shown in the Maya Forest waterlands, eco-borderlands may be abundant in resources yet scarce for their inhabitants. Neighboring communities may also present contrasting realities only because their conditions related remoteness and rurality toward their own respective national centers are distinct. In other words, neighboring communities are located on different sides of national borders.

The Maya Forest waterlands are the remaining Maya/forest/water/lands of Mesoamerica. These are lands and resources defended by those who self-identify as contemporary Mayas and as Indigenous with historical roots. These are also forest waterlands of people not included in the "Maya", where the inhabitants of the Maya Forest continue to encounter the modern nation-state, with its borders and policies. These are also deeply entangled with forest waterlands: The pending questions of ownership, title, rights, and use are embedded in them. Borderlands are encounters between two groups not only in a physical space but over a place. Thus, seemingly uninhabited, remote forest waterlands can well be borderlands for ongoing encounters. In these Maya Forest waterlands, the Indigenous encounter bioprospecting in a challenge over different ownerships of knowledge; the Mayas encounter states in a challenge over different ownerships of forest and subterranean waters; rural peasants encounter tourists in a challenge over different ownerships of livelihoods and housing; forest people encounter archaeologists in a challenge over different ownerships of the ruins and towns; and conservationists entangle with all of these, with their competing mappings and contested discourses.

In this book, we have been inspired by Borderlands Studies – especially borderland historians, who often build on global environmental history. They have managed to shift our analytical angles toward other regionalities defined by other borders. Often, these angles have traditionally pertained to our analytical edges: Oceans, plains, the Arctic. Yet, when the maps are turned and centered otherwise, we suddenly obtain a different understanding of the world. We gain insight on how these socioenvironmental edges played an active part in creating the world as we know it. Of course, historians have the benefit of employing

Image 6.2 After Shipwrecking: A Different Glimpse of the Forest Waterlands of the "Maya Canyon"
Source: Hanna Laako, 2018

these distinct angles as their research timelines are not yet necessarily dependent on modern international borders. Instead, we – political scientists – are often methodologically bound to these borders and frames as compulsory contemporary realities. We are also bound to the national academic systems and funding, which tend to restrict research to national boundaries and problem solving. We are all easily subjected to methodological nationalism.

However, our finding is that the Maya Forest, with its ecoregions, is also an invitation to rethink and extend beyond this methodological nationalism. It is an invitation to *rethink transboundarity*, just as our shipwrecking experience in the rocky forest waterlands of the Usumacinta River imposed on us. While we cannot dismiss the existence of international borders, given that they continue to dictate so much of our sociopolitical world, it does not mean that our sociopolitical world would be automatically and solely bound to them. As we have shown in this book, borders are fluid and shapeshift in time, space, and place as they are human made. It goes without saying that the contemporary climate crisis and biodiversity loss obligate us to extend beyond these bordered lands, waterlines, and forest walls. Diversity flourishes in their interstices.

Index

Note: Information in figures and tables is indicated by page numbers in *italics* and **bold**, respectively.

For Product Safety Concerns and Information please contact our EU
representative GPSR@taylorandfrancis.com
Taylor & Francis Verlag GmbH, Kaufingerstraße 24, 80331 München, Germany

www.ingramcontent.com/pod-product-compliance
Lightning Source LLC
LaVergne TN
LVHW021123300926
849160LV00010B/844

* 9 7 8 1 0 3 2 5 5 1 1 1 1 *